SpringerBriefs in Mathematics

SpringerBriefs in Mathematics showcases expositions in all areas of mathematics and applied mathematics. Manuscripts presenting new results or a single new result in a classical field, new field, or an emerging topic, applications, or bridges between new results and already published works, are encouraged. The series is intended for mathematicians and applied mathematicians.

Titles from this series are indexed by Web of Science, Mathematical Reviews, and zbMATH.

More information about this series at http://www.springer.com/series/10030

Osamu Fujino

Iitaka Conjecture

An Introduction

 Springer

Osamu Fujino
Osaka, Japan

ISSN 2191-8198   ISSN 2191-8201   (electronic)
SpringerBriefs in Mathematics
ISBN 978-981-15-3346-4   ISBN 978-981-15-3347-1   (eBook)
https://doi.org/10.1007/978-981-15-3347-1

Mathematics Subject Classification (2010): Primary 14D06, Secondary 14E30

This Springer imprint is published by the registered company Springer Nature Singapore Pte Ltd.
The registered company address is: 152 Beach Road, #21-01/04 Gateway East, Singapore 189721, Singapore

To my parents, Yuzo and Michiko

Preface

The ambitious program for the birational classification of higher-dimensional algebraic varieties discussed by Shigeru Iitaka in [12] is usually called the Iitaka program. Now it is known that the heart of the Iitaka program is the Iitaka conjecture. The main purpose of this book is to make the Iitaka conjecture more accessible. This book is not an introductory textbook for the Iitaka program but is an introduction to the Iitaka conjecture.

Let X be a smooth projective variety defined over $\mathbb{C}$. We put

$$\kappa(X) = \limsup_{m \to \infty} \frac{\log \dim_{\mathbb{C}} H^0(X, \mathcal{O}_X(mK_X))}{\log m}$$

if $H^0(X, \mathcal{O}_X(lK_X)) \neq 0$ for some positive integer l, where K_X is the canonical divisor of X. If $H^0(X, \mathcal{O}_X(lK_X)) = 0$ for every positive integer l, then we put $\kappa(X) = -\infty$. We call $\kappa(X)$ the Kodaira dimension of X and know that it is a very important birational invariant of X. We can check that

$$\kappa(X) \in \{-\infty, 0, 1, \ldots, \dim X\}.$$

In [11], Shigeru Iitaka defined $\kappa(X)$ and introduced the Iitaka fibration, which is a generalization of the notion of elliptic surfaces. Then he reached the following conjecture.

Conjecture (Iitaka Conjecture C) Let $f : X \to Y$ be a surjective morphism between smooth projective varieties with connected fibers. Then the inequality

$$\kappa(X) \geq \kappa(X_y) + \kappa(Y)$$

holds, where X_y is a sufficiently general fiber of $f : X \to Y$.

This conjecture first appeared in [12] and is still an important open problem. More generally, we have:

Conjecture (Iitaka Conjecture $\overline{C}$) Let $f : X \rightarrow Y$ be a dominant morphism between algebraic varieties whose general fibers are irreducible. Then we have the following inequality:

$$\overline{\kappa}(X) \geq \overline{\kappa}(Y) + \overline{\kappa}(X_y),$$

where X_y is a sufficiently general fiber of $f : X \rightarrow Y$.

Note that $\overline{\kappa}(X)$ denotes the logarithmic Kodaira dimension of X, which was also defined by Shigeru Iitaka in [13] in order to extend his framework, that is, the Iitaka program, for noncomplete algebraic varieties. We note that $\kappa(X) = \overline{\kappa}(X)$ holds when X is a smooth projective variety. Therefore, Conjecture C is a special case of Conjecture $\overline{C}$.

Around 1980, Shigefumi Mori initiated his theory of extremal rays (see [Mo1]). Then the minimal model program (MMP, for short) soon became the standard theory for the birational classification of higher-dimensional algebraic varieties. In some sense, it superseded the Iitaka program. Roughly speaking, we have already known that the above conjectures follow from the good minimal model conjecture (see [Kaw6], [N4], [F13], and [Has2]). We note that the good minimal model conjecture is the main goal of the minimal model program. For the details of the minimal model program, see [F12].

Theorem *Assume that the good minimal model conjecture holds true. Then Conjectures C and $\overline{C}$ also hold true.*

Or, more generally, we have:

Theorem *Conjectures C and $\overline{C}$ follow from the generalized abundance conjecture.*

The above two theorems strongly support Conjectures C and $\overline{C}$ because the experts believe that the good minimal model conjecture undoubtedly holds true. We note that the good minimal model conjecture is still widely open, although the minimal model program has developed drastically during the last decade.

Conjecture C was generalized as follows.

Conjecture (Generalized Iitaka Conjecture C^+) Let $f : X \rightarrow Y$ be a surjective morphism between smooth projective varieties with connected fibers. Assume that $\kappa(Y) \geq 0$. Then the inequality

$$\kappa(X) \geq \kappa(X_y) + \max\{\mathrm{Var}(f), \kappa(Y)\}$$

holds, where X_y is a sufficiently general fiber of $f : X \rightarrow Y$ and $\mathrm{Var}(f)$ denotes Viehweg's variation of $f : X \rightarrow Y$.

Eckart Viehweg posed the following conjecture and proved that Conjecture C^+ follows from Conjecture Q.

Conjecture (Viehweg Conjecture Q) Let $f : X \to Y$ be a surjective morphism between smooth projective varieties with connected fibers such that $\kappa(X_{\bar{\eta}}) \geq 0$, where $X_{\bar{\eta}}$ is the geometric generic fiber of $f : X \to Y$. Assume that $\mathrm{Var}(f) = \dim Y$, where $\mathrm{Var}(f)$ denotes Viehweg's variation of $f : X \to Y$. Then $f_* \omega_{X/Y}^{\otimes k}$ is big in the sense of Viehweg for some positive integer k.

In this book, we describe Viehweg's theory of weakly positive sheaves and big sheaves and show that Conjecture C^+ follows from Conjecture Q. Then we prove the Iitaka Conjecture C in the following cases:

- the base space Y is of general type (see [Vi3]),
- the geometric generic fiber is of general type (see [Kol3]), and
- general fibers are elliptic curves (see [Vi1]).

Finally, we prove the Iitaka Conjecture $\overline{C}$ under the assumption:

- $\dim X - \dim Y = 1$ (see [Kaw2]).

All the above cases are well known to the experts. However, the proof is not so easy to access for the younger generation. The author hopes that this book will make the Iitaka Conjectures C and $\overline{C}$ more accessible. Our choice of topics is biased and reflects the author's taste. In this book, we do not treat the Hodge theoretic part or the complex analytic part of the Iitaka Conjectures C and $\overline{C}$. We will simplify and generalize known arguments and some results with the aid of the weak semistable reduction theorem by Abramovich–Karu (see [Abk]) and the existence of relative canonical models for fiber spaces whose geometric generic fiber is of general type by Birkar–Cascini–Hacon–McKernan (see [BCHM]).

Osaka, Japan   Osamu Fujino

Acknowledgements

The author was partially supported by JSPS KAKENHI Grant Numbers JP24684002, JP24224001, JP16H06337, JP16H03925 during the preparation of this book. He thanks Tetsushi Ito for useful discussions. He also thanks Takeshi Abe and Kaoru Sano for answering his questions. The original version of [F1] was written in 2003 in Princeton. The author is grateful to the Institute for Advanced Study for its hospitality. He was partially supported by a grant from the National Science Foundation: DMS-0111298. He would like to thank Professor Noboru Nakayama for comments on [F1] and Professor Kalle Karu for sending him [Kar1]. He thanks Jinsong Xu for pointing out a mistake in a preliminary version of this book. In 2018, he gave a series of lectures based on a draft of this book in Osaka. He thanks Mayu Adachi, Yuki Kaneko, Keisuke Miyamoto, Yusuke Suyama, and Daisuke Takebayashi, who attended those lectures. He thanks Sung Rak Choi, Sho Ejiri, and Kenta Hashizume for reading a draft and giving him useful comments. He also thanks Yoshinori Gongyo for supporting him on many occasions. Finally, he would like to thank Masayuki Nakamura of Springer-Verlag, who kindly and patiently waited for the completion of the manuscript.

Last but not least, he would like to thank Yoko, Kentaro, and Makoto.

December 2019   Osamu Fujino

Contents

Chapter 1
Overview

1.1 What is the Iitaka Program?

Let us start with a quick review of the *Iitaka program* in the 1970s for the reader's convenience. The basic references for this section are [I2], [U1], and [Ft2] (see also [Es], [I6], and [U2]).

Let X be a smooth projective variety defined over $\mathbb{C}$. We put

$$\kappa(X) = \begin{cases} \max_{m \in \mathbb{Z}_{>0}} \{\dim \Phi_{|mK_X|}(X)\} & \text{if } |mK_X| \neq \emptyset \text{ for some } m \in \mathbb{Z}_{>0}, \\ -\infty & \text{otherwise}, \end{cases}$$

where $\Phi_{|mK_X|}(X)$ denotes the closure of the image of the rational map

$$\Phi_{|mK_X|} : X \dashrightarrow \mathbb{P}^{\dim|mK_X|},$$

and call $\kappa(X)$ the *Kodaira dimension* of X. We note that $\kappa(X)$ is a birational invariant of X. By definition, we have

$$\kappa(X) \in \{-\infty, 0, 1, \ldots, \dim X\}.$$

The following definition is due to Iitaka.

Definition 1.1.1 (*Varieties of hyperbolic type, parabolic type, and elliptic type*) A smooth projective variety X is called a variety of *hyperbolic type*, *parabolic type*, and *elliptic type* if $\kappa(X) = \dim X$, $\kappa(X) = 0$, and $\kappa(X) = -\infty$, respectively.

When $\kappa(X) > 0$, Iitaka proved that the rational maps

$$\Phi_{|mK_X|} : X \dashrightarrow Y_m = \Phi_{|mK_X|}(X)$$

© The Author(s), under exclusive license to Springer Nature Singapore Pte Ltd. 2020
O. Fujino, *Iitaka Conjecture*, SpringerBriefs in Mathematics,
https://doi.org/10.1007/978-981-15-3347-1_1

for all sufficiently large integers m with $|mK_X| \neq \emptyset$ are birationally equivalent to a fixed surjective morphism

$$\Phi_\infty : X_\infty \to Y_\infty$$

between smooth projective varieties with connected fibers such that $\kappa(F) = 0$ holds, where F is a sufficiently general fiber of $\Phi_\infty : X_\infty \to Y_\infty$. We call $\Phi_\infty : X_\infty \to Y_\infty$ the *Iitaka fibration* of X. It is a generalization of the notion of elliptic surfaces. By taking Iitaka fibrations, we can reduce the birational classification of higher-dimensional smooth projective varieties to:

- the study of smooth projective varieties of hyperbolic, parabolic, and elliptic type, and
- the study of fiber spaces whose sufficiently general fibers are of parabolic type.

In [I2, Sect. 4], Iitaka explained his ideas and 12 interesting open problems. Although Iitaka discussed the bimeromorphic classification of compact complex manifolds in [I2, Sect. 5], we will treat only algebraic varieties in this book.

Problem 1.1.2 (*cf.* [I2, Problem 6]) Let $f : X \to Y$ be a surjective morphism between smooth projective varieties with connected fibers and let F be a sufficiently general fiber of $f : X \to Y$. Assume that $\kappa(F) = \dim F$ and $\kappa(Y) = \dim Y$ hold. Then does the equality $\kappa(X) = \dim X$ hold true?

Problem 1.1.3 (*cf.* [I2, Problem 8]) Let $f : X \to Y$ be a surjective morphism between smooth projective varieties with connected fibers and let F be a sufficiently general fiber of $f : X \to Y$. Assume that $\kappa(F) = 0$ and $\kappa(Y) = \dim Y$ hold. Then does the equality $\kappa(X) = \dim Y$ hold true?

In [I2, Footnote 45)], Iitaka posed his famous Conjecture C and said that Problems 1.1.2 and 1.1.3 are special cases of Conjecture C.

Conjecture 1.1.4 (Iitaka Conjecture C) *Let $f : X \to Y$ be a surjective morphism between smooth projective varieties with connected fibers. Then the inequality*

$$\kappa(X) \geq \kappa(X_y) + \kappa(Y)$$

holds, where X_y is a sufficiently general fiber of $f : X \to Y$.

Note that the questions in Problems 1.1.2 and 1.1.3 are now known to be true (see, for example, Theorems 1.2.9 and 1.2.12 below), although Conjecture C is still open.

Remark 1.1.5 (cf. [U2]) The original Iitaka Conjecture C was formulated for compact complex manifolds in [I2]. In Ueno's supplementary footnote 7 in [I2], it was pointed out that Conjecture C does not always hold true for compact complex manifolds. The reader can find an explicit example in [U1, Remark 15.3]. Kenji Ueno said that the importance of Conjecture C had not been fully understood when Iitaka proposed many conjectures in [I2]. The Iitaka Conjecture C has become one of the central issues in the various attempts to solve the several problems in [I2].

Let X be a smooth projective variety of parabolic type or elliptic type. If $q(X) = \dim_{\mathbb{C}} H^0(X, \mathcal{O}_X) > 0$, then we take the *Albanese mapping*

$$\alpha : X \to \text{Alb}(X).$$

When X is a surface of parabolic type (resp. elliptic type), X is birationally equivalent to a hyperelliptic surface or an Abelian surface (resp. a nonrational ruled surface) by the Enriques–Kodaira classification of smooth projective surfaces. For the study of $\alpha : X \to \text{Alb}(X)$, Conjecture C is very important. Ueno posed the following conjecture closely related to Conjecture C.

Conjecture 1.1.6 (Ueno Conjecture K) *If X is a smooth projective variety of parabolic type, then the Albanese mapping $\alpha : X \to \text{Alb}(X)$ is surjective with connected fibers and it is birational to an étale fiber bundle over $\text{Alb}(X)$ which is trivialized by a finite étale base change.*

We note that the Iitaka program does not say anything about smooth projective varieties X with $\kappa(X) = -\infty$ and $q(X) = 0$.

Then Takao Fujita, Eckart Viehweg, Yujiro Kawamata, János Kollár, and others attacked the above conjectures and various related problems. In this book, we will explain some of their work in detail. We note that Conjecture K is still open. Now we know that Conjectures C and K follow from the good minimal model conjecture (see Conjecture 2.3.60 below) by Kawamata's result. For the details, see [Kaw6] (see also Sect. 4.5.2). Thus the minimal model program, which was initiated by Shigefumi Mori around 1980, superseded the Iitaka program in some sense.

1.2 The Main Results

In this section, we explain what we will prove in this book. We will work over $\mathbb{C}$, the complex number field, throughout this section.

Let us recall $\overline{C}_{n,n-1}$, that is, the subadditivity theorem of the logarithmic Kodaira dimension for morphisms of relative dimension one, which is the main result of [Kaw2]. Note that [Kaw2] is Kawamata's master thesis to the Faculty of Science, University of Tokyo. More precisely, Kawamata's master thesis consists of [Kaw1] and [Kaw2].

Theorem 1.2.1 ([Kaw2, Theorem 1] and Theorem 5.2.1) *Let $f : X \to Y$ be a dominant morphism of algebraic varieties. We assume that the general fiber $X_y = f^{-1}(y)$ is an irreducible curve. Then we have the following inequality for logarithmic Kodaira dimensions:*
$$\overline{\kappa}(X) \geq \overline{\kappa}(Y) + \overline{\kappa}(X_y).$$

Note that Theorem 1.2.1 plays an important role in [F9]. One of the main purposes of this book is to make Theorem 1.2.1 more accessible. We give a proof of Theorem

1.2.1 without depending on Kawamata's original paper [Kaw2]. In general, we have the following conjecture.

Conjecture 1.2.2 (Subadditivity of logarithmic Kodaira dimension, Iitaka Conjecture $\overline{C}$) *Let $f : X \to Y$ be a dominant morphism between algebraic varieties whose general fibers are irreducible. Then we have the following inequality:*

$$\overline{\kappa}(X) \geq \overline{\kappa}(Y) + \overline{\kappa}(X_y),$$

where X_y is a sufficiently general fiber of $f : X \to Y$.

Theorem 1.2.1 says that Conjecture 1.2.2 holds true when $\dim X - \dim Y = 1$. Conjecture 1.2.2 is usually called Conjecture $\overline{C}_{n,m}$ when $\dim X = n$ and $\dim Y = m$. Thus, Theorem 1.2.1 means that $\overline{C}_{n,n-1}$ is true. We note the following theorem, which is one of the main consequences of [F13] and [F18] (see also [Has2]).

Theorem 1.2.3 (Lemma 2.3.36 and Theorem 5.3.1) *Conjecture 1.2.2 follows from the generalized abundance conjecture.*

The generalized abundance conjecture (see Conjecture 2.3.59) is one of the most important and difficult problems in the theory of minimal models and is still widely open. For the details, see [F12] and [F13]. Theorem 1.2.3 strongly supports Conjecture 1.2.2.

Let us recall various conjectures related to Conjecture 1.2.2. Obviously, Conjecture 1.2.2 is a generalization of the famous Iitaka Conjecture C.

Conjecture 1.2.4 (Iitaka Conjecture C) *Let $f : X \to Y$ be a surjective morphism between smooth projective varieties with connected fibers. Then the inequality*

$$\kappa(X) \geq \kappa(X_y) + \kappa(Y)$$

holds, where X_y is a sufficiently general fiber of $f : X \to Y$.

The following more precise conjecture is due to Viehweg.

Conjecture 1.2.5 (Generalized Iitaka Conjecture C^+) *Let $f : X \to Y$ be a surjective morphism between smooth projective varieties with connected fibers. Assume that $\kappa(Y) \geq 0$. Then the inequality*

$$\kappa(X) \geq \kappa(X_y) + \max\{\mathrm{Var}(f), \kappa(Y)\}$$

holds, where X_y is a sufficiently general fiber of $f : X \to Y$ and $\mathrm{Var}(f)$ denotes Viehweg's variation of $f : X \to Y$.

In Sect. 4.1, we describe that the generalized Iitaka conjecture (see Conjecture 1.2.5) follows from Viehweg's Conjecture Q (see Conjecture 1.2.6 below). For this purpose, we treat the basic properties of weakly positive sheaves and big sheaves, and Viehweg's base change trick in Sect. 3.1. Almost everything in Sect. 3.1 is contained in Viehweg's papers [Vi3] and [Vi4]. Moreover, we discuss Viehweg's very important arguments for direct images of relative pluricanonical bundles and adjoint bundles, which are also contained in Viehweg's papers [Vi3] and [Vi4], with some refinements in Sect. 3.3. Our treatment in Sect. 4.1 is essentially the same as Viehweg's original one (see [Vi3, Sect. 7]). However, it is slightly simplified and refined by the use of the weak semistable reduction theorem due to Abramovich–Karu (see [AbK] and Sect. 2.3.4).

We note that Viehweg's Conjecture Q is as follows:

Conjecture 1.2.6 (Viehweg Conjecture Q) *Let $f : X \to Y$ be a surjective morphism between smooth projective varieties with connected fibers such that $\kappa(X_{\overline{\eta}}) \geq 0$, where $X_{\overline{\eta}}$ is the geometric generic fiber of $f : X \to Y$. Assume that $\mathrm{Var}(f) = \dim Y$, where $\mathrm{Var}(f)$ denotes Viehweg's variation of $f : X \to Y$. Then $f_*\omega_{X/Y}^{\otimes k}$ is big in the sense of Viehweg for some positive integer k.*

If $\dim X = n$ and $\dim Y = m$ in the above conjectures, then Conjectures C, C^+, and Q are usually called Conjectures $C_{n,m}$, $C_{n,m}^+$, and $Q_{n,m}$ respectively.

In [Kaw6], Kawamata proves Conjecture 1.2.6 under the assumption that the geometric generic fiber of $f : X \to Y$ has a good minimal model (see [Kaw6, Theorem 1.1] and Sect. 4.5.2). Note that [Kaw6], which is a generalization of Viehweg's paper [Vi4], treats infinitesimal Torelli problems for the proof of Conjecture 1.2.6. In this book, we do not discuss infinitesimal Torelli problems or the results in [Kaw6].

In Sect. 4.3, we give a relatively simple proof of Viehweg's Conjecture Q (see Conjecture 1.2.6) under the assumption that the geometric generic fiber of $f : X \to Y$ is of general type. The main theorem of Sect. 4.3, that is, Theorem 4.3.1, is better than the well-known results by Kollár [Kol3] and Viehweg [Vi6] for fiber spaces whose geometric generic fiber is of general type.

Theorem 1.2.7 (Theorem 4.3.1 and Remarks 4.3.2 and 4.3.3) *Let $f : X \to Y$ be a surjective morphism between smooth projective varieties with connected fibers. Assume that the geometric generic fiber $X_{\overline{\eta}}$ of $f : X \to Y$ is of general type and that $\mathrm{Var}(f) = \dim Y$. Then there exists a generically finite surjective morphism $\tau : Y' \to Y$ from a smooth projective variety Y' such that $f'_*\omega_{X'/Y'}^{\otimes k}$ is a nef and big locally free sheaf on Y' for some positive integer k, where X' is a resolution of the main component of $X \times_Y Y'$ and $f' : X' \to Y'$ is the induced morphism.*

We do not need the theory of variations of (mixed) Hodge structure for the proof of Theorem 1.2.7. One of the main new ingredients of Theorem 1.2.7 (see Theorem 4.3.1) is the effective freeness due to Popa–Schnell (see [PopS]).

Theorem 1.2.8 (Theorem 3.2.1) *Let $f : X \to Y$ be a surjective morphism from a smooth projective variety X to a projective variety Y with $\dim Y = n$. Let k be a* *positive integer and let $\mathcal{L}$ be an ample invertible sheaf on Y such that $|\mathcal{L}|$ is free. Then we have*

$$H^i(Y, f_*\omega_X^{\otimes k} \otimes \mathcal{L}^{\otimes l}) = 0$$

for every $i > 0$ and every $l \geq nk + k - n$. By Castelnuovo–Mumford regularity, $f_*\omega_X^{\otimes k} \otimes \mathcal{L}^{\otimes l}$ is generated by global sections for every $l \geq k(n + 1)$.

We prove this effective freeness in Sect. 3.2 for the reader's convenience (see Theorem 3.2.1). The proof of Theorem 3.2.1 is a clever application of a generalization of Kollár's vanishing theorem and is very simple. Another important ingredient of the proof of Theorem 1.2.7 is the existence of relative canonical models for fiber spaces whose geometric generic fiber is of general type by Birkar–Cascini–Hacon–McKernan (see [BCHM]). Thus, we have:

Theorem 1.2.9 (…, Kollár, Viehweg, …) *Let $f : X \to Y$ be a surjective morphism between smooth projective varieties with connected fibers whose geometric generic fiber is of general type. Assume that $\kappa(Y) \geq 0$. Then we have*

$$\kappa(X) \geq \kappa(X_y) + \max\{\mathrm{Var}(f), \kappa(Y)\}$$
$$= \dim X - \dim Y + \max\{\mathrm{Var}(f), \kappa(Y)\},$$

where X_y is a sufficiently general fiber of $f : X \to Y$.

Theorem 1.2.9 completely solves Problem 1.1.2.

In Sect. 4.4, we quickly review elliptic fibrations and see that Conjecture 1.2.6 holds for elliptic fibrations. Therefore, we have:

Theorem 1.2.10 (Viehweg, …) *Let $f : X \to Y$ be a surjective morphism between smooth projective varieties with connected fibers whose general fibers are elliptic curves. Assume that $\kappa(Y) \geq 0$. Then we have*

$$\kappa(X) \geq \kappa(X_y) + \max\{\mathrm{Var}(f), \kappa(Y)\}$$
$$= \max\{\mathrm{Var}(f), \kappa(Y)\},$$

where X_y is a sufficiently general fiber of $f : X \to Y$.

By combining Theorem 1.2.9 with Theorem 1.2.10, we have:

Corollary 1.2.11 (Viehweg, see [Vi1]) *Let $f : X \to Y$ be a surjective morphism between smooth projective varieties whose general fibers are irreducible curves. Then we have*

$$\kappa(X) \geq \kappa(X_y) + \kappa(Y),$$

where X_y is a general fiber of $f : X \to Y$.

Note that the proof of Theorem 1.2.1 in Sect. 5.2 uses Theorems 1.2.9 and 1.2.10. More precisely, we use the solution of Conjecture 1.2.6 for morphisms of relative dimension one. We also note that Kawamata's original proof of Theorem 1.2.1 heavily depends on Viehweg's paper [Vi1]. We do not directly use [Vi1] in this book. Therefore, the reader can understand the proof of Theorem 1.2.1 in this book without referring to [Vi1].

Finally, this book is also an introduction to Viehweg's theory of weakly positive sheaves and big sheaves. In Sect. 4.2, we quickly prove the following theorem as an easy direct application of Viehweg's weak positivity theorem for direct images of relative pluricanonical bundles.

Theorem 1.2.12 ([Vi3, Corollary IV]) *Let* $f : X \to Y$ *be a surjective morphism between smooth projective varieties with connected fibers. Assume that Y is of general type, that is, $\kappa(Y) = \dim Y$. Then we have*

$$\kappa(X) = \kappa(X_y) + \kappa(Y)$$
$$= \kappa(X_y) + \dim Y,$$

where X_y is a sufficiently general fiber of $f : X \to Y$.

Theorem 1.2.12 is a complete solution of Problems 1.1.2 and 1.1.3.

As we stated above, some of Viehweg's arguments in [Vi3] and [Vi4] are simplified by the use of the weak semistable reduction theorem due to Abramovich and Karu. We hope that this book will make Viehweg's ideas in [Vi3] and [Vi4] more accessible.

1.3 Historical Note

This book is a revised version of the author's preprint:

- Osamu Fujino, Subadditivity of the logarithmic Kodaira dimension for morphisms of relative dimension one revisited, preprint 2014,

which was written in Kyoto. The above preprint is a completely revised and expanded version of the author's unpublished short note [F1]:

- Osamu Fujino, $\overline{C}_{n,n-1}$ revisited, preprint 2003,

which was written when the author was staying at the Institute for Advanced Study, Princeton. In Princeton, he had enough time to study mathematics and write various notes. The main purpose of [F1] was to understand Kawamata's paper [Kaw2], which did not seem to be so easy to access for the younger generation. The paper [Kaw2] seems to contain some mistakes (see Sect. 5.1 below). To the best knowledge of the author, there is no rigorous proof of $\overline{C}_{n,n-1}$ in the literature, although the experts can prove $\overline{C}_{n,n-1}$ with some efforts. The author hopes that this book will make the Iitaka conjecture and $\overline{C}_{n,n-1}$ more accessible.

1.4 Precise Contents

We now describe the contents of each chapter more precisely.

Chapter 2: We collect some basic definitions and results. In Sect. 2.1, we fix the notations. In Sect. 2.2, we explain the prerequisites for this book. In Sect. 2.3, we quickly explain the Iitaka dimension, the Kodaira dimension, the logarithmic Kodaira dimension, various cyclic covering tricks, the weak semistable reduction theorem due to Abramovich–Karu, Kollár's torsion-freeness and vanishing theorem, and so on. The weak semistable reduction theorem due to Abramovich–Karu explained in this chapter plays an important role in this book.

Chapter 3: We explain the basic properties of Viehweg's weakly positive sheaves and big sheaves. Section 3.1 collects the basic properties of nef locally free sheaves, weakly positive sheaves, and big sheaves. We also explain Viehweg's base change trick. Section 3.2 is devoted to the effective freeness due to Popa and Schnell. It is a relatively new result. Although it is very powerful, the proof is surprisingly easy. In Sect. 3.3, we discuss the weak positivity of relative (pluri-)canonical bundles and adjoint bundles following Viehweg's original treatment. In this section, we slightly simplify some of Viehweg's arguments. In Sect. 3.4, we quickly see some further developments without proof. We explain singular hermitian metrics on torsion-free coherent sheaves by Păun–Takayama, Nakayama's theory of ω-sheaves, a twisted weak positivity and an effective generation by Dutta–Murayama.

Chapter 4: In Sect. 4.1, we explain that the generalized Iitaka Conjecture C^+ follows from the Viehweg Conjecture Q. Our treatment may look slightly different from the traditional one because we use the weak semistable reduction theorem by Abramovich–Karu. In Sect. 4.2, we quickly prove the Iitaka Conjecture C under the assumption that the base space is of general type. This case is an easy consequence of Viehweg's weak positivity of direct images of relative pluricanonical bundles. Section 4.3 is one of the main parts of this book. We prove the Viehweg Conjecture Q under the assumption that the geometric generic fiber is of general type. Our proof needs no deep results coming from the theory of variations of Hodge structure. Moreover, we do not need Viehweg's theory of weakly positive sheaves for the proof of Theorem 4.3.1, which is the main result of Sect. 4.3. Our result in this section is sharper than the known results. We note that we use the existence of relative canonical models for fiber spaces whose geometric generic fiber is of general type by Birkar–Cascini–Hacon–McKernan in this section. In Sect. 4.4, we prove the Viehweg Conjecture Q for elliptic fibrations. This section is not self-contained. We use some well-known results on elliptic fibrations and moduli spaces of elliptic curves without proof. In Sect. 4.5, we quickly see some other known cases of the Iitaka conjecture without proof. We treat fiber spaces whose base space has maximal Albanese dimension, fiber spaces whose geometric generic fiber has a good minimal model, and the Iitaka conjecture in positive characteristic.

Chapter 5: In Sect. 5.1, we see the background of $\overline{C}_{n,n-1}$ and explain the author's motivation. Section 5.2 is devoted to the proof of $\overline{C}_{n,n-1}$. This section is one of the main parts of this book and is a completely revised version of the author's unpublished short note written in 2003 in Princeton. Our approach is different from Kawamata's original one in [Kaw2]. In Sect. 5.3, we see some related results on the Iitaka Conjecture $\overline{C}$. We explain that the Iitaka Conjecture $\overline{C}$ follows from the generalized abundance conjecture. We also explain some new results on the Iitaka Conjecture $\overline{C}$ by Kovács–Patakfalvi and Hashizume. In Sect. 5.4, which is an appendix, we give a quick proof of Kawamata's vanishing statement in [Kaw2] for the reader's convenience, although we do not need it in this book.

Chapter 6: This chapter contains appendices. In Sect. 6.1, we give a simple proof of $C_{2,1}$ based on the Enriques–Kodaira classification of smooth projective surfaces for the reader's convenience. In Sect. 6.2, we show that Iitaka's Conjecture C_n follows from weaker conjectures: Conjectures $(\geq)_n$ and $(>)_n$. Although we do not use the results in Sect. 6.2 explicitly in this book, the arguments are useful for the study of the Iitaka conjecture.

1.5 References

The lecture note [U1] by Ueno on the Iitaka program, which is now a classic work, is still the best introductory textbook on this topic. It treats not only algebraic varieties but also complex analytic spaces. Iitaka's lecture note [I4] is also a good introductory textbook. It mainly explains the theory of the logarithmic Kodaira dimension due to Iitaka himself. The lecture note [Popp] by Popp explains some moduli problems and applications to the Iitaka Conjecture C. The last two chapters of Iitaka's book [I5] treat the Iitaka program. The explanations in the Japanese original version of [I5] are more exciting than those in [I5]. We can find a gentle introduction to the Iitaka dimension and the Iitaka fibration in [Laz1, Sect. 2.1]. Nakayama's book [N4] contains his theory of various analogues of the Kodaira dimension. In [N4], Nakayama treats $\mathbb{R}$-divisors and very sophisticated generalizations of Viehweg's arguments. Therefore, it is not so easy to read. In [BPV] (not [BHPV]), the Iitaka Conjecture $C_{2,1}$ is proved directly and is used for the proof of the Enriques–Kodaira classification of smooth projective surfaces. On the other hand, the proof of the Enriques–Kodaira classification of smooth projective surfaces in [BHPV] does not use $C_{2,1}$ but some ideas of the minimal model program. The change of the proof of the Enriques–Kodaira classification of smooth projective surfaces between [BPV] and [BHPV] may be evidence that the minimal model program superseded the Iitaka program.

Mori's survey article [Mo2] is the only one that explains various results on Conjectures C, C^+, and Q mainly due to Kawamata, Kollár, and Viehweg. We note that Mori's treatment of the theory of the Iitaka dimension in [Mo2] is very sophisticated. To the best knowledge of the author, there are no other survey articles or textbooks on Conjectures C, C^+, and Q. This is one of the author's motivations to write this book. Although [Mo2] is a survey article, it contains some original results. Note that [Mo2, Sect. 5, Part II] treats a prototype of Mori's canonical bundle formula. It becomes the so-called Fujino–Mori canonical bundle formula in [FM].

A relatively new survey article [HPS] by Hacon–Popa–Schnell is a very accessible account of the Iitaka Conjecture C for fiber spaces whose base space has maximal Albanese dimension. We strongly recommend the reader to see [HPS], where the precise version of the Ohsawa–Takegoshi L^2-extension theorem plays a crucial role.

Finally, the reader can find Iitaka's original ideas and Fujita's ideas on the birational classification of higher-dimensional algebraic varieties in [I2] and [Ft2], respectively. Unfortunately, both [I2] and [Ft2] were written in Japanese and have not been translated into English.

Chapter 2
Preliminaries

In this chapter, we recall some fundamental facts and fix the notations in this book.

2.1 Notations, Terminology, and Conventions

Unless otherwise explicitly stated to the contrary, the following conventions will be in force throughout this book.

- A *scheme* is a separated scheme of finite type over $\mathbb{C}$, the complex number field. A *variety* is an integral scheme.
- The words *invertible sheaf* (resp. *locally free sheaf*) and *line bundle* (resp. *vector bundle*) are used interchangeably.
- The set of integers (resp. rational numbers or real numbers) is denoted by $\mathbb{Z}$ (resp. $\mathbb{Q}$ or $\mathbb{R}$). The set of positive integers (resp. nonnegative integers) is denoted by $\mathbb{Z}_{>0}$ (resp. $\mathbb{Z}_{\geq 0}$).

2.2 Prerequisites

We assume that the reader is familiar with basic algebraic geometry, at the level of [Har3] and [Laz1]. It is highly desirable that the reader is familiar with the basic results of the Iitaka program, at the level of [U1, Chaps. II and III].

© The Author(s), under exclusive license to Springer Nature Singapore Pte Ltd. 2020
O. Fujino, *Iitaka Conjecture*, SpringerBriefs in Mathematics,
https://doi.org/10.1007/978-981-15-3347-1_2

2.3 Preliminary Results

In this section, we collect some basic notations and results for the reader's convenience. For the details, see [U1], [KolM], [Mo2], [Laz1], [F7], [F12], and so on.

2.3.1 Basic Definitions

Let us recall some basic definitions.

2.3.1 (Simple normal crossing divisors) Let X be a smooth variety. An effective Cartier divisor D on X is said to be a *simple normal crossing divisor* if for each closed point p of X, a local defining equation f of D at p can be written as

$$f = z_1 \cdots z_{j_p}$$

in $\mathcal{O}_{X,p}$, where $\{z_1, \ldots, z_{j_p}\}$ is a part of a regular system of parameters.

2.3.2 (Generic generation) Let $\mathcal{F}$ be a coherent sheaf on a smooth quasi-projective variety X. We say that $\mathcal{F}$ is *generated by global sections over U*, where U is a Zariski open set of X, if the natural map

$$H^0(X, \mathcal{F}) \otimes \mathcal{O}_X \to \mathcal{F}$$

is surjective over U. We say that $\mathcal{F}$ is *generically generated by global sections* if $\mathcal{F}$ is generated by global sections over some nonempty Zariski open set of X.

2.3.3 (Operations for coherent sheaves) Let $\mathcal{F}$ be a coherent sheaf on a normal variety X. We put

$$\mathcal{F}^* = \mathcal{H}om_{\mathcal{O}_X}(\mathcal{F}, \mathcal{O}_X)$$

and

$$\mathcal{F}^{**} = (\mathcal{F}^*)^*.$$

If the natural map

$$\mathcal{F} \to \mathcal{F}^{**}$$

is an isomorphism, then $\mathcal{F}$ is called a *reflexive* sheaf. For the basic properties of reflexive sheaves, we recommend the reader to see [Har4, Sect. 1]. We put

$$\widehat{S^\alpha}(\mathcal{F}) = (S^\alpha(\mathcal{F}))^{**}$$

for every positive integer α, where $S^\alpha(\mathcal{F})$ is the αth symmetric product of $\mathcal{F}$, and

$$\widehat{\det}(\mathcal{F}) = (\wedge^r \mathcal{F})^{**},$$

where $r = \mathrm{rank}\mathcal{F}$. When X is smooth, $\widehat{\det}(\mathcal{F})$ is invertible since it is a reflexive sheaf of rank one.

We note the following definition of exceptional divisors.

2.3.4 (Exceptional divisors) Let $f : X \to Y$ be a proper surjective morphism between normal varieties. Let E be a Weil divisor on X. We say that E is f-*exceptional* if $\mathrm{codim}_Y f(\mathrm{Supp} E) \geq 2$. Note that f is not always assumed to be birational. When $f : X \to Y$ is a birational morphism, $\mathrm{Exc}(f)$ denotes the exceptional locus of f.

2.3.5 (Sufficiently general fibers) Let $f : X \to Y$ be a morphism between varieties. Then a *sufficiently general fiber* F of $f : X \to Y$ means that $F = f^{-1}(y)$, where y is any point contained in a countable intersection of nonempty Zariski open sets of Y. A sufficiently general fiber is sometimes called a *very general fiber* in the literature. A *sufficiently general point* y of a variety Y is any closed point contained in a countable intersection of nonempty Zariski open sets of Y.

Remark 2.3.6 Let V be a variety defined over $\mathbb{C}$ and let U_m be a nonempty Zariski open set of V for every $m \in \mathbb{Z}$. Then V is a locally compact Hausdorff space and U_m is a dense open set of V for every $m \in \mathbb{Z}$ in the classical topology. Therefore, by Baire's category theorem,

$$U = \bigcap_{m \in \mathbb{Z}} U_m$$

is dense in V in the classical topology. Note that U is not always open in V.

We sometimes use $\mathbb{Q}$-divisors in this book. Fortunately, we do not need $\mathbb{R}$-divisors, which play a crucial role in the recent developments of the minimal model program (see [BCHM] and [F12]), in this book.

2.3.7 (Operations for $\mathbb{Q}$-divisors) Let $D = \sum_i a_i D_i$ be a $\mathbb{Q}$-divisor on a normal variety X, where D_i is a prime divisor on X for every i, $D_i \neq D_j$ for $i \neq j$, and $a_i \in \mathbb{Q}$ for every i. Then we put

$$\lfloor D \rfloor = \sum_i \lfloor a_i \rfloor D_i, \quad \{D\} = D - \lfloor D \rfloor, \quad \text{and} \quad \lceil D \rceil = -\lfloor -D \rfloor.$$

Note that $\lfloor a_i \rfloor$ is the integer which satisfies $a_i - 1 < \lfloor a_i \rfloor \leq a_i$. We also note that $\lfloor D \rfloor$, $\lceil D \rceil$, and $\{D\}$ are called the *round-down*, *round-up*, and *fractional part* of D respectively.

2.3.8 ($\mathbb{Q}$-linear equivalence) Let B_1 and B_2 be two $\mathbb{Q}$-divisors on a normal variety X. Then B_1 is $\mathbb{Q}$-*linearly equivalent* to B_2, denoted by $B_1 \sim_{\mathbb{Q}} B_2$, if $mB_1 \sim mB_2$ for some positive integer m. We note that $\sim$ denotes the *linear equivalence* of two divisors.

2.3.9 ($\mathbb{Q}$-Cartier divisors) Let D be a $\mathbb{Q}$-divisor on a normal variety X. If there exists some positive integer m such that mD is a Cartier divisor, then we say that D is a $\mathbb{Q}$-*Cartier* divisor.

2.3.10 (Nef line bundles) A line bundle $\mathcal{L}$ on a projective variety X is called *nef* if $\mathcal{L} \cdot C \geq 0$ for every projective curve C on X. A $\mathbb{Q}$-Cartier divisor D on X is called *nef* if $\mathcal{O}_X(mD)$ is a nef line bundle for some positive integer m such that mD is Cartier.

2.3.11 (Canonical sheaves and canonical divisors) Let X be an equidimensional scheme of dimension n and let $\omega_X^\bullet$ be the dualizing complex of X. Then we put

$$\omega_X := \mathcal{H}^{-n}(\omega_X^\bullet)$$

and call it the *canonical sheaf* of X.

We further assume that X is normal. Then a *canonical divisor* K_X of X is a Weil divisor on X such that

$$\mathcal{O}_{X_{\mathrm{sm}}}(K_X) \simeq \Omega_{X_{\mathrm{sm}}}^n$$

holds, where X_{sm} is the largest smooth Zariski open set of X. Note that K_X is a well-defined Weil divisor on X up to linear equivalence.

It is well known that

$$\mathcal{O}_X(K_X) \simeq \omega_X$$

holds when X is normal.

If $f : X \to Y$ is a morphism between Gorenstein schemes, then we put

$$\omega_{X/Y} := \omega_X \otimes f^* \omega_Y^{\otimes -1}.$$

If $f : X \to Y$ is a morphism from a normal scheme X to a normal Gorenstein scheme Y, then we put

$$K_{X/Y} := K_X - f^* K_Y.$$

Let us quickly explain the definition of singularities of pairs for the reader's convenience. For the details of singularities of pairs, see [F6], [F7], and [F12].

2.3.12 (Singularities of pairs)   Let X be a normal variety and let Δ be an effective $\mathbb{Q}$-divisor on X such that $K_X + \Delta$ is $\mathbb{Q}$-Cartier. Let $f : Y \to X$ be a resolution of singularities. We write

$$K_Y = f^*(K_X + \Delta) + \sum_i a_i E_i$$

such that $f_* \left(\sum_i a_i E_i \right) = -\Delta$, and $a(E_i, X, \Delta) = a_i$. Note that the *discrepancy* $a(E, X, \Delta) \in \mathbb{Q}$ can be defined for every prime divisor E *over* X. If $a(E, X, \Delta) > -1$ for every exceptional divisor E over X, then (X, Δ) is called a *plt* pair. If $a(E, X, \Delta) > -1$ (resp. ≥ -1) for every divisor E over X, then (X, Δ) is called a *klt* (resp. an *lc*) pair. In this book, if $\Delta = 0$ and $a(E, X, 0) \geq 0$ for every divisor E over X, then we say that X has only *canonical singularities*.

We note the following well-known result. We will use it in Chap. 4.

Lemma 2.3.13 *Let X be a Gorenstein variety. Then X has only canonical singularities if and only if X has only rational singularities.*

We give a Proof of Lemma 2.3.13 for the reader's convenience.

Proof It is well known that X has only rational singularities if it has only canonical singularities (see, for example, [F12, Sects. 3.13, 3.14, and 3.15]). If X has only rational singularities, then $f_*\omega_Y \simeq \omega_X$ holds for every resolution of singularities $f : Y \to X$ (see, for example, [F12, Lemma 3.12.2]). Note that $a(E, X, 0) \in \mathbb{Z}$ since X is Gorenstein. Therefore, $a(E, X, 0) \geq 0$ holds for every divisor E over X by $f_*\omega_Y \simeq \omega_X$. This means that X has only canonical singularities. $\qquad\square$

2.3.14 (Horizontal and vertical divisors) Let $f : X \to Y$ be a dominant morphism between normal varieties and let D be a $\mathbb{Q}$-divisor on X. We can write

$$D = D_{\text{hor}} + D_{\text{ver}}$$

such that every irreducible component of D_{hor} (resp. D_{ver}) is mapped (resp. not mapped) onto Y. If $D = D_{\text{hor}}$ (resp. $D = D_{\text{ver}}$), D is said to be *horizontal* (resp. *vertical*).

2.3.2 Iitaka Dimension and Kodaira Dimension

Let us quickly see Iitaka's theory of $\kappa(X, D)$.

2.3.15 (Iitaka dimension and Kodaira dimension) Let D be a Cartier divisor on a normal projective variety X. The *Iitaka dimension* $\kappa(X, D)$ is defined as follows:

$$\kappa(X, D) = \begin{cases} \max_{m \in \mathbb{Z}_{>0}} \{\dim \Phi_{|mD|}(X)\} & \text{if } |mD| \neq \emptyset \text{ for some } m \in \mathbb{Z}_{>0}, \\ -\infty & \text{otherwise}, \end{cases}$$

where $\Phi_{|mD|} : X \dashrightarrow \mathbb{P}^{\dim|mD|}$ and $\Phi_{|mD|}(X)$ denotes the closure of the image of the rational map $\Phi_{|mD|}$. Of course,

$$\kappa(X, D) \in \{-\infty, 0, 1, \ldots, \dim X\}$$

holds. Let D be a $\mathbb{Q}$-Cartier divisor on X. Then we put

$$\kappa(X, D) = \kappa(X, m_0 D),$$

where m_0 is a positive integer such that $m_0 D$ is Cartier. Let $\mathcal{L}$ be an invertible sheaf on X. Then we can define $\kappa(X, \mathcal{L})$ analogously.

Let X be a smooth projective variety. Then we put

$$\kappa(X) = \kappa(X, K_X).$$

Note that $\kappa(X)$ is usually called the *Kodaira dimension* of X. If X is an arbitrary projective variety, then we put

$$\kappa(X) = \kappa(\widetilde{X}, K_{\widetilde{X}}),$$

where $\widetilde{X} \to X$ is a projective birational morphism from a smooth projective variety $\widetilde{X}$. It is easy to see that $\kappa(X)$ is independent of the choice of $\widetilde{X} \to X$.

Lemma 2.3.16 *Let $f : X \dashrightarrow X'$ be a birational map between smooth projective varieties. Then there exists a natural $\mathbb{C}$-linear isomorphism*

$$f^* : H^0(X', \mathcal{O}_{X'}(mK_{X'})) \xrightarrow{\sim} H^0(X, \mathcal{O}_X(mK_X))$$

for every nonnegative integer m. In particular, $\kappa(X) = \kappa(X')$ holds, that is, $\kappa(X)$ is a birational invariant of smooth projective varieties.

Proof We can take a closed subset Σ of X such that $f : X \setminus \Sigma \to X'$ is a morphism with $\mathrm{codim}_X \Sigma \geq 2$. We take $\omega \in H^0(X', \mathcal{O}_{X'}(mK_{X'}))$. Then we can directly check that

$$f^*\omega \in H^0(X \setminus \Sigma, \mathcal{O}_X(mK_X)) = H^0(X, \mathcal{O}_X(mK_X))$$

since $\mathrm{codim}_X \Sigma \geq 2$, By considering $f^{-1} : X' \dashrightarrow X$, we see that

$$f^* : H^0(X', \mathcal{O}_{X'}(mK_{X'})) \to H^0(X, \mathcal{O}_X(mK_X))$$

is a natural $\mathbb{C}$-linear isomorphism. $\qquad\qquad\qquad\qquad\qquad\qquad\qquad\square$

Remark 2.3.17 Let X be a normal projective variety and let D be a Cartier divisor on X. Then it is well known that

$$\kappa(X, D) = \limsup_{m \to \infty} \frac{\log \dim_{\mathbb{C}} H^0(X, \mathcal{O}_X(mD))}{\log m} \tag{2.3.1}$$

holds. Therefore, we can adopt (2.3.1) as a definition of $\kappa(X, D)$. For the details, see, for example, [Laz2, Corollary 2.1.38] and [U1, Theorem 8.1] (see also [U1, Remark 8.3]). Or, we recommend the reader to see Iitaka's original definition of κ in [I1].

Let us quickly review some more details for the reader's convenience. We put

$$\mathbb{N}(X, D) = \{l \in \mathbb{Z}_{>0} \mid H^0(X, \mathcal{O}_X(lD)) \neq 0\}.$$

Note that $\mathbb{N}(X, D) = \emptyset$ if and only if $\kappa(X, D) = -\infty$. If $\mathbb{N}(X, D) \neq \emptyset$, then we can find positive integers e and l_0 such that

$$\mathbb{N}(X, D) \cap \{l \in \mathbb{Z} \mid l \geq l_0\} = e\mathbb{Z} \cap \{l \in \mathbb{Z} \mid l \geq l_0\}.$$

Note that e is sometimes called the *exponent* of $\mathbb{N}(X, D)$. In this situation, we can take positive rational numbers α, β, and a positive integer m_0 such that the following inequalities:

$$\alpha m^{\kappa(X,D)} \leq \dim_{\mathbb{C}} H^0(X, \mathcal{O}_X(meD)) \leq \beta m^{\kappa(X,D)} \tag{2.3.2}$$

hold for every integer $m \geq m_0$. In [I1], Iitaka originally used (2.3.2) to define $\kappa(X, D)$.

Remark 2.3.18 Let X be a smooth projective variety. We put

$$R(X) = \bigoplus_{m \geq 0} H^0(X, \mathcal{O}_X(mK_X))$$

and call it the *canonical ring* of X. By Lemma 2.3.16, $R(X)$ is a birational invariant of X. By [BCHM], we know that $R(X)$ is a finitely generated $\mathbb{C}$-algebra. By definition, $R(X) = \mathbb{C}$ if and only if $\kappa(X) = -\infty$. When $\kappa(X) \geq 0$, we put

$$X_{\mathrm{can}} = \mathrm{Proj}\, R(X)$$

and usually call it the *canonical model* of X. It is easy to see that

$$\kappa(X) = \dim X_{\mathrm{can}}$$

holds since X_{can} is isomorphic to $\Phi_{|mK_X|}(X)$, which is the closure of the image of the rational map $\Phi_{|mK_X|} : X \dashrightarrow \mathbb{P}^{\dim|mK_X|}$, for some sufficiently large and divisible positive integer m.

Remark 2.3.19 The theory of $\kappa(X, D)$ and $\kappa(X)$ works over any algebraically closed field k of characteristic zero. However, if $\#k$ is countable, then $U = \bigcap_{m \in \mathbb{Z}} U_m$, where U_m is a nonempty Zariski open set for every $m \in \mathbb{Z}$, may be empty (see Remark 2.3.6). Therefore, when $k \neq \mathbb{C}$, it is reasonable to replace X_y, a sufficiently general fiber, with $X_{\overline{\eta}}$, the geometric generic fiber, in Conjecture C, and so on (see Conjecture 1.2.4, and so on). We note that the equality $\kappa(X_y) = \kappa(X_{\overline{\eta}})$ holds when $k = \mathbb{C}$ (see Lemma 2.3.28 below).

Let us see easy examples.

Example 2.3.20 Let C be a smooth projective curve and let $g(C)$ denote the genus of C, that is,

$$g(C) = \dim_{\mathbb{C}} H^0(C, \mathcal{O}_C(K_C)) = \dim_{\mathbb{C}} H^1(C, \mathcal{O}_C).$$

Then we have

$$\kappa(C) = \begin{cases} -\infty & \text{if } g(C) = 0, \text{ equivalently, } C = \mathbb{P}^1, \\ 0 & \text{if } g(C) = 1, \text{ that is, } C \text{ is an elliptic curve,} \\ 1 & \text{if } g(C) \geq 2. \end{cases}$$

Example 2.3.21 Let X be an n-dimensional smooth hypersurface in $\mathbb{P}^{n+1}$ with $n \geq 1$. We note that

$$\omega_X \simeq (\omega_{\mathbb{P}^{n+1}} \otimes \mathcal{O}_{\mathbb{P}^{n+1}}(X))|_X$$

holds by adjunction and that $\omega_{\mathbb{P}^{n+1}} = \mathcal{O}_{\mathbb{P}^{n+1}}(-n-2)$. Therefore, we have

$$\kappa(X) = \begin{cases} -\infty & \text{if } \deg X \leq n+1, \\ 0 & \text{if } \deg X = n+2, \\ n = \dim X & \text{if } \deg X \geq n+3. \end{cases}$$

Example 2.3.22 Let X_1 and X_2 be smooth projective varieties. Then

$$\kappa(X_1 \times X_2) = \kappa(X_1) + \kappa(X_2)$$

holds. For the details, see [U1, Example 6.6, 3)].

Example 2.3.23 Let $f : X \to Y$ be a generically finite surjective morphism between smooth projective varieties. Then the inequality

$$\kappa(X) \geq \kappa(Y)$$

holds. This is because we can write $K_X = f^* K_Y + R$ for some effective divisor R on X.

The following lemma is very important. We will use it repeatedly without mentioning it explicitly.

Lemma 2.3.24 *Let* $f : X \to Y$ *be a finite surjective morphism between normal projective varieties and let* D *be a* $\mathbb{Q}$*-Cartier divisor on* Y*. Then the equality*

$$\kappa(X, f^* D) = \kappa(Y, D)$$

holds.

For the Proof of Lemma 2.3.24, see, for example, [U1, Theorem 5.13].

Definition 2.3.25 (*Big* ($\mathbb{Q}$-)*divisors*) Let D be a $\mathbb{Q}$-Cartier divisor on a normal projective variety X. If $\kappa(X, D) = \dim X$, then we say that D is *big*.

Let us recall the definition of varieties of general type. A smooth projective variety of general type was called a variety of hyperbolic type in the Iitaka program (see Definition 1.1.1).

Definition 2.3.26 (*Varieties of general type*) Let X be a smooth projective variety. We say that X is *of general type* when K_X is big.

We recall an important lemma due to Kodaira on big divisors.

Lemma 2.3.27 (Kodaira) *Let D be a Cartier divisor on a normal projective variety X and let H be any ample divisor on X. If $\kappa(X, D) = \dim X$, that is, D is big, then there exists a positive integer a such that $aD - H$ is linearly equivalent to an effective Cartier divisor, equivalently,*

$$H^0(X, \mathcal{O}_X(aD - H)) \neq 0.$$

For the Proof of Lemma 2.3.27, see, for example, [F12, Lemma 2.1.21].

In this book, we will repeatedly use the following lemma, which is an easy consequence of the semicontinuity theorem, without mentioning it explicitly.

Lemma 2.3.28 *Let $f : X \to Y$ be a projective surjective morphism between normal varieties with connected fibers and let D be a Cartier divisor on X. Then we can take U, which is a countable intersection of nonempty Zariski open sets of Y, such that*

$$\dim_{\mathbb{C}} H^0(X_y, \mathcal{O}_X(mD)|_{X_y}) = \dim_{\overline{\mathbb{C}(Y)}} H^0(X_{\overline{\eta}}, \mathcal{O}_X(mD)|_{X_{\overline{\eta}}})$$

holds for every $y \in U$ and every $m \in \mathbb{Z}$, where $X_{\overline{\eta}}$ is the geometric generic fiber of $f : X \to Y$ and $\overline{\mathbb{C}(Y)}$ is the algebraic closure of $\mathbb{C}(Y)$, the function field of Y. In particular,

$$\kappa(X_{\overline{\eta}}, \mathcal{O}_X(D)|_{X_{\overline{\eta}}}) = \kappa(X_y, \mathcal{O}_X(D)|_{X_y})$$

holds for every $y \in U$.

Proof Without loss of generality, we may assume that $f : X \to Y$ is flat and $X_y = X \times_Y \operatorname{Spec}\mathbb{C}(y)$ is normal for every $y \in Y$ by shrinking Y suitably (see [G, Théorème (12.2.4)]). By the semicontinuity theorem (see, for example, [Har3, Chap. III, Theorem 12.8]) and the flat base change theorem (see, for example, [Har3, Chap. III, Proposition 9.3]), we can take a nonempty Zariski open set U_m of Y such that

$$\dim_{\mathbb{C}} H^0(X_y, \mathcal{O}_X(mD)|_{X_y}) = \dim_{\overline{\mathbb{C}(Y)}} H^0(X_{\overline{\eta}}, \mathcal{O}_X(mD)|_{X_{\overline{\eta}}})$$

holds for $y \in U_m$. We put

$$U = \bigcap_{m \in \mathbb{Z}} U_m.$$

Then, this U is a desired subset of Y. By Remark 2.3.17, we have $\kappa(X_{\overline{\eta}}, \mathcal{O}_X(D)|_{X_{\overline{\eta}}}) = \kappa(X_y, \mathcal{O}_X(D)|_{X_y})$ for every $y \in U$. $\qquad\qquad\square$

Remark 2.3.29 In Lemma 2.3.28, let U' be a Zariski open set of Y such that f is flat over U' and that X_y is normal for every $y \in U'$. Then, by the Proof of Lemma 2.3.28, we have

$$\kappa(X_{\overline{\eta}}, \mathcal{O}_X(D)|_{X_{\overline{\eta}}}) \leq \kappa(X_y, \mathcal{O}_X(D)|_{X_y})$$

for every $y \in U'$. This is because

$$\dim_{\overline{\mathbb{C}(Y)}} H^0(X_{\overline{\eta}}, \mathcal{O}_X(mD)|_{X_{\overline{\eta}}}) \leq \dim_{\mathbb{C}} H^0(X_y, \mathcal{O}_X(mD)|_{X_y})$$

holds for every $m \in \mathbb{Z}_{>0}$ and $y \in U'$ by the semicontinuity theorem.

Theorem 2.3.30 (Iitaka fibrations) *Let X be a normal projective variety and let D be a Cartier divisor on X such that $\kappa(X, D) > 0$. Then for all sufficiently large integers k with $H^0(X, \mathcal{O}_X(kD)) \neq 0$, the rational maps $\Phi_{|kD|} : X \dashrightarrow Y_k$ are birationally equivalent to a fixed surjective morphism $\Phi_\infty : X_\infty \to Y_\infty$ between smooth projective varieties with connected fibers, and the restriction of D to a sufficiently general fiber of Φ_∞ has Iitaka dimension zero. More specifically, there exists a commutative diagram*

$$
\begin{array}{ccc}
X & \xleftarrow{\ u_\infty\ } & X_\infty \\
\Phi_{|kD|} \downarrow & & \downarrow \Phi_\infty \\
Y_k & \xleftarrow{\ v_k\ } & Y_\infty
\end{array}
$$

of rational maps and morphisms for every sufficiently large positive integer k with $H^0(X, \mathcal{O}_X(kD)) \neq 0$, where the horizontal maps are birational and u_∞ is a morphism. We have $\dim Y_\infty = \kappa(X, D)$. Moreover, if we put $D_\infty = u_\infty^* D$, and take $F \subset X_\infty$ to be a sufficiently general fiber of Φ_∞, then

$$\kappa(F, D_\infty|_F) = 0.$$

The morphism $\Phi_\infty : X_\infty \to Y_\infty$ is called the Iitaka fibration *associated to D. It is unique up to birational equivalence. When $\Phi_{|kD|} : X \dashrightarrow \Phi_{|kD|}(X) \subset \mathbb{P}^N$ is birationally equivalent to the Iitaka fibration $\Phi_\infty : X_\infty \to Y_\infty$, we simply say that $\Phi_{|kD|} : X \dashrightarrow \mathbb{P}^N$ gives an Iitaka fibration.*

This theorem is well known. For the details, see [Laz1, Theorem 2.1.33 and Definition 2.1.34] and [U1, Chap. III].

The following inequality due to Iitaka is well known and is easy to check.

Lemma 2.3.31 (Easy addition) *Let $f : X \to Y$ be a surjective morphism between normal projective varieties with connected fibers and let D be a $\mathbb{Q}$-Cartier divisor on X. Then we have*

$$\kappa(X, D) \leq \dim Y + \kappa(X_y, D_y),$$

where X_y is a general fiber of $f : X \to Y$ and $D_y = D|_{X_y}$.

Proof We may assume that D is Cartier by replacing D with mD for some suitable positive integer m. If $\kappa(X, D) = -\infty$, then the inequality is obvious. We can easily check the desired inequality when $\kappa(X, D) = 0$. Therefore, we may assume that $\kappa(X, D) > 0$. We take a large and divisible positive integer m such that $\Phi_{|mD|}$: $X \dashrightarrow \mathbb{P}^N$ gives an Iitaka fibration. We consider the following diagram:

$$
\begin{array}{ccc}
X & \overset{\varphi}{\dashrightarrow} & \mathbb{P}^N \times Y \xrightarrow{\ p_1\ } \mathbb{P}^N \\
\ \ {\scriptstyle f} \downarrow & \swarrow {\scriptstyle p_2} & \\
Y & &
\end{array}
$$

where $\varphi = \Phi_{|mD|} \times f$ and p_1 and p_2 are natural projections. Let Z be the image of φ in $\mathbb{P}^N \times Y$. Then we obtain that

$$
\begin{aligned}
\kappa(X, D) &= \dim p_1(Z) \\
&\leq \dim Z \\
&= \dim Y + \dim Z_y \\
&\leq \dim Y + \kappa(X_y, D|_{X_y}),
\end{aligned}
$$

where y is a general point of Y. This is the desired inequality. $\qquad\square$

Remark 2.3.32 In Lemma 2.3.31, let U' be a Zariski open set of Y such that f is flat over U' and that X_y is normal for every $y \in U'$. Let m_0 be a positive integer such that $m_0 D$ is Cartier. Then, by Lemma 2.3.28, Remark 2.3.29, and the Proof of Lemma 2.3.31, we have

$$
\kappa(X, D) \leq \dim Y + \kappa(X_{\overline{\eta}}, \mathcal{O}_X(m_0 D)|_{X_{\overline{\eta}}}) \leq \dim Y + \kappa(X_y, \mathcal{O}_X(m_0 D)|_{X_y})
$$

for every $y \in U'$, where $X_{\overline{\eta}}$ is the geometric generic fiber of $f : X \to Y$.

2.3.33 (Logarithmic Kodaira dimension) Let V be a variety. By Nagata's theorem, we have a complete variety $\overline{V}$ which contains V as a dense Zariski open set. By Hironaka's resolution of singularities, we have a smooth projective variety $\overline{Z}$ and a projective birational morphism $\mu : \overline{Z} \to \overline{V}$ such that if $Z = \mu^{-1}(V)$, then $\overline{D} = \overline{Z} - Z = \mu^{-1}(\overline{V} - V)$ is a simple normal crossing divisor on $\overline{Z}$. The *logarithmic Kodaira dimension* $\overline{\kappa}(V)$ of V is defined as

$$
\overline{\kappa}(V) = \kappa(\overline{Z}, K_{\overline{Z}} + \overline{D}),
$$

where κ denotes the Iitaka dimension in 2.3.15. Note that $\overline{\kappa}(V)$ is well defined, that is, $\overline{\kappa}(V)$ is independent of the choice of $(\overline{Z}, \overline{D})$. We can check it by Hironaka's resolution of singularities and Lemma 2.3.34 below.

Lemma 2.3.34 *Let* $f : X \to Y$ *be a birational morphism between smooth projective varieties. Let* D_X *and* D_Y *be simple normal crossing divisors on* X *and* Y, *respectively. Assume that* $\operatorname{Supp} f^* D_Y \subset \operatorname{Supp} D_X$, *that is,* $f(X \setminus D_X) \subset Y \setminus D_Y$. *Then there exists an effective divisor* E *on* X *such that*

$$K_X + D_X = f^*(K_Y + D_Y) + E \qquad\qquad (2.3.3)$$

holds. Therefore, the pull-back

$$f^* : H^0(Y, \mathcal{O}_Y(m(K_Y + D_Y))) \to H^0(X, \mathcal{O}_X(m(K_X + D_X)))$$

is an inclusion for every nonnegative integer m.

We further assume that $f_* D_X = D_Y$ holds. Then E in (2.3.3) is f-exceptional. In this situation, the pull-back

$$f^* : H^0(Y, \mathcal{O}_Y(m(K_Y + D_Y))) \to H^0(X, \mathcal{O}_X(m(K_X + D_X)))$$

is an isomorphism for every nonnegative integer m. In particular, the equality

$$\kappa(X, K_X + D_X) = \kappa(Y, K_Y + D_Y)$$

holds.

Proof By direct local calculations, we can check that $K_X + D_X \geq f^*(K_Y + D_Y)$ holds. Therefore, we have an effective divisor E on X such that

$$K_X + D_X = f^*(K_Y + D_Y) + E$$

holds. Thus we obtain that the pull-back

$$f^* : H^0(Y, \mathcal{O}_Y(m(K_Y + D_Y))) \to H^0(X, \mathcal{O}_X(m(K_X + D_X)))$$

is an inclusion for every nonnegative integer m.

If we further assume that $f_* D_X = D_Y$ holds, then it is obvious that E is f-exceptional. In this case, we have

$$f_* \mathcal{O}_X(m(K_X + D_X)) \simeq \mathcal{O}_Y(m(K_Y + D_Y))$$

for every nonnegative integer m. This implies that the pull-back

$$f^* : H^0(Y, \mathcal{O}_Y(m(K_Y + D_Y))) \to H^0(X, \mathcal{O}_X(m(K_X + D_X)))$$

is an isomorphism for every nonnegative integer m and that

$$\kappa(X, K_X + D_X) = \kappa(Y, K_Y + D_Y)$$

holds.   □

We note the following easy but important example.

Example 2.3.35 Let C be a (not necessarily complete) smooth curve. Then we can easily see that

$$\overline{\kappa}(C) = \begin{cases} -\infty & \text{if } C = \mathbb{P}^1 \text{ or } \mathbb{A}^1, \\ 0 & \text{if } C \text{ is an elliptic curve or } \mathbb{G}_m, \\ 1 & \text{otherwise.} \end{cases}$$

We explicitly state an easy lemma for the reader's convenience.

Lemma 2.3.36 Let $g : V \to W$ be a dominant morphism between varieties whose general fibers are irreducible. Then we can construct a commutative diagram:

$$\begin{array}{ccc} V & \xleftarrow{\ p\ } & X \\ {\scriptstyle g}\downarrow & & \downarrow{\scriptstyle f} \\ W & \xleftarrow{\ q\ } & Y \end{array}$$

with the following properties:

(i) $f : X \to Y$ is a surjective morphism between smooth projective varieties with connected fibers.

(ii) There exist simple normal crossing divisors D_X and D_Y on X and Y, respectively.

(iii) $p : X \setminus D_X \dashrightarrow V$ and $q : Y \setminus D_Y \dashrightarrow W$ are projective birational morphisms.

(iv) $\operatorname{Supp} f^* D_Y \subset \operatorname{Supp} D_X$ holds.

(v) The following equalities $\overline{\kappa}(V) = \kappa(X, K_X + D_X)$, $\overline{\kappa}(W) = \kappa(Y, K_Y + D_Y)$, and $\overline{\kappa}(V_w) = \kappa(F, K_F + D_X|_F)$ hold, where V_w is a sufficiently general fiber of $g : V \to W$ and F is that of $f : X \to Y$.

Therefore,

$$\overline{\kappa}(V) \geq \overline{\kappa}(W) + \overline{\kappa}(V_w)$$

is equivalent to

$$\kappa(X, K_X + D_X) \geq \kappa(Y, K_Y + D_Y) + \kappa(F, K_F + D_X|_F).$$

Proof By Nagata's theorem, we take completions $\overline{V}$ and $\overline{W}$ of V and W, respectively. By considering the graph of the rational map $\overline{V} \dashrightarrow \overline{W}$ induced by $g : V \to W$, we may assume that $g : V \to W$ extends to $\overline{g} : \overline{V} \to \overline{W}$. Then, by Hironaka's resolution of singularities, we get a commutative diagram with the desired properties. $\square$

The reader can find a different approach to Iitaka dimensions and Iitaka fibrations in [Mo2, Sect. 1]. In some sense, it is much more sophisticated than the usual one.

2.3.3   Cyclic Covers

Here, we collect some useful cyclic covering tricks. The following description of cyclic covers are sometimes very useful.

2.3.37 (Cyclic covers) Let X be a smooth variety and let $\mathcal{L}$ be an invertible sheaf on X. Assume that $\mathcal{L}^{\otimes N} = \mathcal{O}_X(D)$ for some positive integer N, where D is an effective Cartier divisor on X such that $\operatorname{Supp} D$ is a simple normal crossing divisor on X. Let $s \in H^0(X, \mathcal{L}^{\otimes N})$ be a section whose zero divisor is D. Then the dual of

$$s : \mathcal{O}_X \to \mathcal{L}^{\otimes N}$$

defines an $\mathcal{O}_X$-algebra structure on

$$\mathcal{A}' = \bigoplus_{i=0}^{N-1} \mathcal{L}^{\otimes -i}.$$

We put

$$\pi' : Y' = \operatorname{Spec}_X \mathcal{A}' \to X.$$

Let $\pi : Y \to X$ be the finite morphism induced by the normalization $Y \to Y'$ of Y'. We say that Y is the *cyclic cover associated to* $\mathcal{L}^{\otimes N} = \mathcal{O}_X(D)$. Then we can easily see that Y has only quotient singularities. We can explicitly write

$$\pi : Y = \operatorname{Spec}_X \mathcal{A} \to X$$

where

$$\mathcal{A} = \bigoplus_{i=0}^{N-1} \left(\mathcal{L}^{\otimes -i} \otimes \mathcal{O}_X \left(\left\lfloor \frac{iD}{N} \right\rfloor \right) \right).$$

Note that $\mathcal{A}$ is an $\mathcal{O}_X$-algebra and $\mathcal{A}' \subset \mathcal{A}$. For the details, see, for example, [EsV, Sect. 3] and [Vi7, Lemma 2.3].

The following lemmas are very useful covering tricks, which are mainly due to Kawamata (see [Kaw3, Theorem 17 and Corollary 19]).

Lemma 2.3.38 *Let X be a smooth quasi-projective variety and let D be a simple normal crossing divisor on X. Let $D = \sum_{j=1}^r D_j$ be the irreducible decomposition. Let $N_1, \ldots, N_r$ be given positive integers. Then there exist a simple normal crossing divisor Σ on X, a smooth quasi-projective variety Z, and a finite morphism $\tau : Z \to X$ such that:*

(i) $\tau^* D_j = N_j (\tau^* D_j)_{\mathrm{red}}$ for $j = 1, \ldots, r$, and
(ii) $D \leq \Sigma$, τ is étale over $X \setminus \Sigma$, and $\operatorname{Supp} \tau^* \Sigma$ is a simple normal crossing divisor on Z.

Sketch of the Proof of Lemma 2.3.38 We take an ample invertible sheaf $\mathcal{A}$ on X such that $\mathcal{A}^{\otimes N_i} \otimes \mathcal{O}_X(-D_i)$ is generated by global sections. We put $n = \dim X$. We take smooth divisors

$$H_1^{(1)}, \ldots, H_n^{(1)}, H_1^{(2)}, \ldots, H_n^{(2)}, \ldots, H_1^{(r)}, \ldots, H_n^{(r)}$$

in general position with $\mathcal{A}^{\otimes N_i} = \mathcal{O}_X(D_i + H_j^{(i)})$ for $1 \le i \le r$ and $1 \le j \le n$. Let $Z_j^{(i)}$ be the cyclic cover associated to $\mathcal{A}^{\otimes N_i} = \mathcal{O}_X(D_i + H_j^{(i)})$ for $1 \le i \le r$ and $1 \le j \le n$ (see 2.3.37). Let Z be the normalization of

$$\left(Z_1^{(1)} \times_X \times \cdots \times_X Z_n^{(1)} \right) \times_X \cdots \times_X \left(Z_1^{(r)} \times_X \times \cdots \times_X Z_n^{(r)} \right).$$

By construction, we can check that Z is a smooth quasi-projective variety that is finite over X, $\tau : Z \to X$ is étale over $X \setminus \Sigma$ with $\Sigma = D + \sum_{i,j} H_j^{(i)}$, and $\operatorname{Supp}\tau^*\Sigma$ is a simple normal crossing divisor on Z. For the details, we recommend the reader to see [EsV, 3.19. Lemma]. $\square$

Lemma 2.3.39 *Let $f : Y \to X$ be a finite surjective morphism from a normal quasi-projective variety Y to a smooth quasi-projective variety X. Assume that f is étale over $X \setminus \Sigma_Y$, where Σ_Y is a simple normal crossing divisor on X. Then we can take a finite surjective morphism $g : Z \to Y$ from a smooth quasi-projective variety Z such that $f \circ g : Z \to X$ is étale over $X \setminus \Sigma_Z$, where Σ_Z is a simple normal crossing divisor on X such that $\Sigma_Y \le \Sigma_Z$ and that $\operatorname{Supp}(f \circ g)^*\Sigma_Z$ is a simple normal crossing divisor on Z.*

Proof We closely follow [Vi7, Corollary 2.6]. Let $\Sigma_Y = \sum_{j=1}^r D_j$ be the irreducible decomposition. For $j = 1, \ldots, r$, we put

$$N_j = \operatorname*{lcm}_i \{ e(\Delta_j^i) \mid \Delta_j^i \text{ is an irreducible component of } f^{-1}(D_j) \},$$

where $e(\Delta_j^i)$ denotes the ramification index of Δ_j^i over D_j. By Lemma 2.3.38, we can take a simple normal crossing divisor $\Sigma_{\widetilde{X}}$ on X with $\Sigma_Y \le \Sigma_{\widetilde{X}}$ and a finite cover $\tau : \widetilde{X} \to X$ from a smooth quasi-projective variety $\widetilde{X}$ such that $\widetilde{X}$ is étale over $X \setminus \Sigma_{\widetilde{X}}$. Let Z be the normalization of an irreducible component of $Y \times_X \widetilde{X}$.

$$
\begin{array}{ccc}
Y & \xleftarrow{\ g\ } & Z \\
{\scriptstyle f}\downarrow & & \downarrow{\scriptstyle h} \\
X & \xleftarrow{\ \tau\ } & \widetilde{X}
\end{array}
$$

Note that $\tau : \widetilde{X} \to X$ is constructed as a chain of finite cyclic covers. Therefore, $g : Z \to Y$ is also a chain of finite cyclic covers. The ramification index of an irreducible component of $g^{-1}(\Delta_j^i)$ over Δ_j^i is $N_j e(\Delta_j^i)^{-1}$ and the ramification index of an irreducible component of $(f \circ g)^{-1} D_j$ over D_j is given by

$$\frac{N_j}{e(\Delta_j^i)} \cdot e(\Delta_j^i) = N_j.$$

By the construction of $\tau : \widetilde{X} \to X$, this is the ramification index of an irreducible component of $\tau^{-1}(D_j)$ over D_j. Therefore, $h : Z \to \widetilde{X}$ is unramified in codimension one. Since $\widetilde{X}$ is smooth, h is étale. Therefore, Z is a smooth quasi-projective variety. We put $\Sigma_Z = \Sigma_{\widetilde{X}}$. Then it is easy to see that $f \circ g : Z \to X$ is étale over $X \setminus \Sigma_Z$ and that $\mathrm{Supp}(f \circ g)^* \Sigma_Z$ is a simple normal crossing divisor on Z. $\qquad \square$

2.3.4 Weakly Semistable Morphisms

In this book, we will repeatedly use the notion of weakly semistable morphisms due to Abramovich–Karu (see [AbK] and [Kar1]).

Let us quickly recall toroidal embeddings (see [KKMS], [AbK], [Kar1], and so on).

Definition 2.3.40 (*Toroidal embeddings, see* [KKMS] *and* [AbK, Definition 1.2]) Let X be a normal variety and let U_X be a smooth Zariski open set of X. We say that $U_X \subset X$ is a *toroidal embedding* if for every closed point $x \in X$ there exist a toric variety X_σ, a closed point $s \in X_\sigma$, and an isomorphism of complete local $\mathbb{C}$-algebras

$$\widehat{\mathcal{O}}_{X,x} \simeq \widehat{\mathcal{O}}_{X_\sigma,s}$$

such that the ideal of $X \setminus U_X$ maps isomorphically to the ideal of $X_\sigma \setminus T_\sigma$, where T_σ is the dense torus in X_σ.

A pair (X_σ, s) in Definition 2.3.40 is called a *local model* at $x \in X$.

Definition 2.3.41 (*Toroidal morphisms, see* [AbK, Definition 1.3]) A dominant morphism $f : (U_X \subset X) \to (U_Y \subset Y)$ of toroidal embeddings is called *toroidal* if for every closed point $x \in X$ there exist local models (X_σ, s) at $x \in X$, (X_τ, t) at $f(x) \in Y$, and a toric morphism $g : X_\sigma \to X_\tau$ such that the following diagram commutes:

$$
\begin{array}{ccc}
\widehat{\mathcal{O}}_{X,x} & \xrightarrow{\simeq} & \widehat{\mathcal{O}}_{X_\sigma,s} \\
\hat{f}^\sharp \uparrow & & \uparrow \hat{g}^\sharp \\
\widehat{\mathcal{O}}_{Y,f(x)} & \xrightarrow{\simeq} & \widehat{\mathcal{O}}_{X_\tau,t}
\end{array}
$$

where $\hat{f}^\sharp$ and $\hat{g}^\sharp$ are the algebra homomorphisms induced by f and g, respectively.

Let $(U_X \subset X)$ be a toroidal embedding. Then we can write $X \setminus U_X = \bigcup_{i \in I} E_i$ where E_i's are irreducible subvarieties of codimension one. In this book, we will always assume that all the E_i's are normal, that is, $(U_X \subset X)$ is a *toroidal embedding without self-intersection* in the sense of [KKMS, Chap. II, Sect. 1].

2.3.42 (Weakly semistable morphisms) Let $f : X \to Y$ be a projective surjective morphism between quasi-projective varieties. Then $f : X \to Y$ is called *weakly semistable* if:

(i) the varieties X and Y admit toroidal structures $(U_X \subset X)$ and $(U_Y \subset Y)$ with $U_X = f^{-1}(U_Y)$,
(ii) with this structure, the morphism f is toroidal,
(iii) the morphism f is equidimensional,
(iv) all the fibers of the morphism f are reduced, and
(v) Y is smooth.

Note that $(U_X \subset X)$ and $(U_Y \subset Y)$ are toroidal embeddings without self-intersection in the sense of [KKMS, Chap. II, Sect. 1]. We also note that f is automatically flat by (iii) since Y is smooth and X is Cohen–Macaulay (see [Har3, Chap. III, Exercise 10.9] and [AlK, Chap. V, Proposition (3.5)]).

If $f : X \to Y$ is weakly semistable and X is smooth, then we say that $f : X \to Y$ is *semistable*.

We note the following very important result.

Lemma 2.3.43 ([AbK, Lemma 6.1]) *Let $f : X \to Y$ be a weakly semistable morphism. Then X has only rational Gorenstein singularities.*

By Lemma 2.3.13, we know that X has only canonical Gorenstein singularities in Lemma 2.3.43. For the Proof of Lemma 2.3.43, see the original paper [AbK].

Let us recall one of the most important results in [AbK] for the reader's convenience.

Theorem 2.3.44 (Toroidalizations of morphisms, see [AbK, Theorem 2.1]) *Let $f : X \to Y$ be a proper surjective morphism between varieties such that the geometric generic fiber $X_{\overline{\eta}}$ is integral. Let Z be a closed subscheme of X with $Z \subsetneq X$. Then there exists a diagram*

$$
\begin{array}{ccccc}
U_{X'} & \hookrightarrow & X' & \xrightarrow{m_X} & X \\
\downarrow & & \downarrow{\scriptstyle f'} & & \downarrow{\scriptstyle f} \\
U_{Y'} & \hookrightarrow & Y' & \xrightarrow{m_Y} & Y
\end{array}
$$

such that m_X and m_Y are projective birational morphisms, X' and Y' are smooth quasi-projective varieties, the inclusions $U_{X'} \subset X'$ and $U_{Y'} \subset Y'$ are toroidal embeddings without self-intersection with the following properties:

(i) *f' is toroidal with respect to $(U_{X'} \subset X')$ and $(U_{Y'} \subset Y')$,*

(ii) $Z' = m_X^{-1}Z$ is a simple normal crossing divisor on X' with $Z' \subset X' \setminus U_{X'}$, and

(iii) there exists a dense Zariski open set U of X such that m_X is an isomorphism over U, $U_{X'} = m_X^{-1}(U)$, and $U \cap Z = \emptyset$.

Remark 2.3.45 Although (iii) in Theorem 2.3.44 was not treated explicitly in [AbK], we can easily make $f' : X' \to Y'$ satisfy (iii) by slightly modifying the proof of [AbK, Theorem 2.1]. Note that (iii) in Theorem 2.3.44 is very useful for various geometric applications.

From now on, we will freely use the notation in [AbK]. For condition (iii) in Theorem 2.3.44, it is sufficient to prove that there exists a Zariski open set U of X such that $U_{X'} = m_X^{-1}(U)$ and that m_X is an isomorphism on $U_{X'}$ in [AbK, Theorem 2.1]. Precisely speaking, we enlarge Z and may assume that $X \setminus Z$ is a Zariski open set of the original X in [AbK, 2.3], and enlarge Δ suitably in [AbK, 2.5]. Then we can construct $m_X : X' \to X$ such that U is a Zariski open set of X, $U_{X'} = m_X^{-1}(U)$, and $m_X : U_{X'} \to U$ is an isomorphism.

The sharpest result in this direction will be treated in [AbTW] (see also [AdLT, Theorem 4.7]).

Once we obtain Theorem 2.3.44, it is not difficult to prove the main result of [AbK].

Theorem 2.3.46 (Weak semistable reduction, see [AbK, Theorem 0.3 and Remark 4.5]) *Let $X \to Y$ be a proper surjective morphism between varieties such that the geometric generic fiber $X_{\overline{\eta}}$ is integral. Then there exist a generically finite projective surjective morphism $Y' \to Y$ from a smooth quasi-projective variety Y' and a projective birational morphism $X' \to X \times_Y Y'$ from a normal quasi-projective variety X' such that the induced map $X' \to Y'$ is weakly semistable and that X' has only quotient singularities.*

Sketch of the Proof of Theorem 2.3.46 By modifying $f : X \to Y$ birationally, we may assume that $f : X \to Y$ is toroidal by Theorem 2.3.44. By [AbK, Proposition 3.1], we may assume that $U_X = f^{-1}(U_Y)$. Let $\Delta_X \to \Delta_Y$ be the associated morphism of polyhedral complexes. By [AbK, Proposition 4.4 and Remark 4.5], that is, by taking suitable subdivisions of Δ_X and Δ_Y, we can make $f : X \to Y$ satisfy (i), (ii), (iii), (v) in 2.3.42 such that X has only quotient singularities. By taking a suitable finite cover of Y (see Lemma 2.3.38 and [AbK, Sect. 5]), we can construct the desired weakly semistable morphism $X' \to Y'$. For the details, see [AbK] and [Kar1]. $\square$

Remark 2.3.47 Quite recently, Adiprasito, Liu, and Temkin solved the semistable reduction conjecture in full generality in [AdLT]. However, we do not use [AdLT] in this book. Theorem 2.3.46 is sufficient for our applications.

The following lemma is easy but very useful.

Lemma 2.3.48 *Let $(U_Y \subset Y)$ be a toroidal embedding. Let $f : X \to Y$ and $g : Z \to Y$ be weakly semistable morphisms with respect to $(U_Y \subset Y)$. Then $V = X \times_Y$*

Z has only rational Gorenstein singularities. We consider the following commutative diagram:

$$\begin{array}{ccc} X & \xleftarrow{\ g'\ } & V \\ {\scriptstyle f}\downarrow & & \downarrow{\scriptstyle f'} \\ Y & \xleftarrow{\ g\ } & Z \end{array}$$

Then we have that

$$g'^{*}\omega_{X/Y} = \omega_{V/Z}$$

and

$$g^{*}f_{*}\omega_{X/Y}^{\otimes n} = f'_{*}g'^{*}\omega_{X/Y}^{\otimes n} = f'_{*}\omega_{V/Z}^{\otimes n}$$

for every integer n.

Proof We note that X and Z are Gorenstein by Lemma 2.3.43. By the flat base change theorem [Ve, Theorem 2] (see also [Har2], [Co], and so on), we see that V is Gorenstein and $g'^{*}\omega_{X/Y} = \omega_{V/Z}$. Since f and g are weakly semistable, we see that V is smooth in codimension one. Therefore, V is a normal variety. Since V is local analytically isomorphic to a toric variety, V has only rational singularities. By the flat base change theorem (see [Har3, Chap. III, Proposition 9.3]), we obtain $g^{*}f_{*}\omega_{X/Y}^{\otimes n} = f'_{*}g'^{*}\omega_{X/Y}^{\otimes n}$ for every integer n. $\qquad\square$

We will repeatedly use the following very important and useful result in this book.

Lemma 2.3.49 ([AbK, Lemma 6.2] and [Kar2, Lemma 2.12 (2)]) *Let $f : X \to Y$ be a weakly semistable morphism and let $\tau : Y' \to Y$ be a morphism from a smooth quasi-projective variety Y' such that $\tau^{-1}(Y \setminus U_Y)$ is a simple normal crossing divisor on Y'. We put $X' = X \times_Y Y'$. Let $\rho : X' \to X$ and $f' : X' \to Y'$ be the projections.*

$$\begin{array}{ccc} X' & \xrightarrow{\ \rho\ } & X \\ {\scriptstyle f'}\downarrow & & \downarrow{\scriptstyle f} \\ Y' & \xrightarrow{\ \tau\ } & Y \end{array}$$

We set $U_{Y'} = \tau^{-1}(U_Y)$ and $U_{X'} = \rho^{-1}(U_X)$. Then $(U_{Y'} \subset Y')$ and $(U_{X'} \subset X')$ are toroidal embeddings without self-intersection, and $f' : X' \to Y'$ is weakly semistable.

For the Proof of Lemma 2.3.49, see [AbK, Lemma 6.2]. It is interesting and important to understand how to construct local toric models of $(U_{X'} \subset X')$.

We note that some results in [Kar1] are slightly better than those in [AbK]. We also note that [Kar1, Chap. 2, Sect. 9] takes care of horizontal divisors. We recommend the interested reader to see [Kar1]. As we said in Remarks 2.3.45 and 2.3.47, [AdLT] and [AbTW] treat much sharper results.

2.3.5 Kollár's Torsion-Freeness and Vanishing Theorem

We need the following generalization of Kollár's torsion-freeness and vanishing theorem in this book.

Theorem 2.3.50 *Let $f : X \to Y$ be a projective surjective morphism between quasi-projective varieties such that X is smooth. Let Δ be an effective $\mathbb{Q}$-divisor on X such that $\mathrm{Supp}\Delta$ is a simple normal crossing divisor and that $\lfloor \Delta \rfloor = 0$. Let D be a Cartier divisor on X.*

(i) (Torsion-freeness). *If $D - (K_X + \Delta)$ is f-semi-ample, then $R^q f_* \mathcal{O}_X(D)$ is torsion-free for every q.*

(ii) (Vanishing theorem). *We further assume that X and Y are both projective. If $D - (K_X + \Delta) \sim_{\mathbb{Q}} f^* H$ for some ample $\mathbb{Q}$-divisor H on Y, then $H^p(Y, R^q f_* \mathcal{O}_X (D)) = 0$ for every $p > 0$ and every q.*

We note that Theorem 2.3.50 is a special case of [Kol5, 10.15 Corollary], [F7, Theorem 6.3], [F12, Theorems 3.16.3 and 5.6.2], and so on. We strongly recommend the interested reader to see [EsV], [Kol5], and [F12, Chaps. 3 and 5] for various vanishing theorems useful for the study of higher-dimensional algebraic varieties. The theory of quasi-log schemes discussed in [F12, Chap. 6] is a very powerful framework for the study of higher-dimensional algebraic varieties by using generalizations of Kollár's torsion-freeness and vanishing theorem.

2.3.6 On Abelian Varieties

We give some supplementary results on Abelian varieties for the reader's convenience (see [F2, Sect. 5. Some remarks on Abelian varieties]). We will use Corollary 2.3.54 only in the Proof of Theorem 1.2.1 in Sect. 5.2.

2.3.51 Let Y be a not necessarily complete variety and let A be an Abelian variety. We put $Z = Y \times A$. Let $\mu : A \times A \to A$ be the multiplication. Then A acts on A naturally by the group law of A. This action induces a natural action on Z. We denote it by $m : Z \times A \to Z$, that is,

$$m : ((y, a), b) \mapsto (y, a + b),$$

where $(y, a) \in Y \times A = Z$ and $b \in A$. Let $p_{1i} : Z \times A \times A \to Z \times A$ be the projection onto the $(1, i)$-factor for $i = 2, 3$, and let $p_{23} : Z \times A \times A \to A \times A$ be the projection onto the $(2, 3)$-factor. Let $p : Z \times A \times A \to Z$ be the first projection and let $p_i : Z \times A \times A \to A$ be the ith projection for $i = 2, 3$. We define the projection $\rho : Z = Y \times A \to A$. We fix a section $s : A \to Z$ such that $s(A) = \{y_0\} \times A$ for a point $y_0 \in Y$. We define morphisms as follows:

$$\pi_i = p_i \circ (s \times id_A \times id_A) \quad \text{for } i = 2, 3$$
$$\pi_{23} = p_{23} \circ (s \times id_A \times id_A), \quad \text{and}$$
$$\pi = \rho \times id_A \times id_A.$$

Let L be an invertible sheaf on Z. We define an invertible sheaf $\mathcal{L}$ on $Z \times A \times A$ as follows:

$$\mathcal{L} = p^* L \otimes (id_Z \times \mu)^* m^* L \otimes (p_{12}^* m^* L)^{\otimes -1} \otimes (p_{13}^* m^* L)^{\otimes -1}$$
$$\otimes \pi^* ((\pi_{23}^* \mu^* s^* L)^{\otimes -1} \otimes \pi_2^* s^* L \otimes \pi_3^* s^* L).$$

Lemma 2.3.52 *Under the above notation, we have*

$$\mathcal{L} \simeq \mathcal{O}_{Z \times A \times A}.$$

Proof It is easy to see that the restrictions $\mathcal{L}$ to $Z \times \{0\} \times A$ and $Z \times A \times \{0\}$ are trivial by the definition of $\mathcal{L}$, where 0 is the origin of A. We can also check that the restriction of $\mathcal{L}$ to $s(A) \times A \times A$ is trivial (see the Proof of [Mu, Sect. 6, Corollary 2]). In particular, $\mathcal{L}|_{\{z_0\} \times A \times A}$ is trivial for any point $z_0 \in s(A) \subset Z$. Therefore, by the theorem of cube (see [Mu, Sect. 6, Theorem]), we obtain that $\mathcal{L}$ is trivial.

We write $T_a = m|_{Z \times \{a\}} : Z \simeq Z \times \{a\} \to Z$, that is,

$$T_a : (y, b) \mapsto (y, b + a),$$

for $(y, b) \in Y \times A = Z$.

Corollary 2.3.53 *By restricting $\mathcal{L}$ to $Z \times \{a\} \times \{b\}$, we obtain*

$$L \otimes T_{a+b}^* L \simeq T_a^* L \otimes T_b^* L,$$

where $a, b \in A$.

As an application of Corollary 2.3.53, we have:

Corollary 2.3.54 *Let D be a Cartier divisor on Z. Then we have*

$$2D \sim T_a^* D + T_{-a}^* D$$

for every $a \in A$. In particular, if Y is complete and D is effective with $D \neq 0$ and is not vertical with respect to $Y \times A \to Y$, then $\kappa(Z, D) > 0$.

Proof We put $L = \mathcal{O}_X(D)$ and $b = -a$. Then we have $2D \sim T_a^* D + T_{-a}^* D$ by Corollary 2.3.53. We assume that $D \neq 0$ and that D is not vertical. Then we have $\text{Supp} D \neq \text{Supp} T_a^* D$ if we choose $a \in A$ suitably. Therefore, $\kappa(X, D) > 0$ if $D \neq 0$ is effective and is not vertical. $\qquad \square$

2.3.7 Invariance of Plurigenera

Although we do not use the following theorem in this book explicitly, we recall it
for the reader's convenience.

Theorem 2.3.55 (Invariance of plurigenera) *Let* $f : X \to Y$ *be a smooth projective
morphism between smooth varieties with connected fibers. Then*

$$\dim_{\mathbb{C}} H^0(X_y, \mathcal{O}_{X_y}(mK_{X_y}))$$

is independent of $y \in Y$ *for every nonnegative integer* m, *where* $X_y = f^{-1}(y)$. *In
particular, the Kodaira dimension* $\kappa(X_y)$ *is independent of* $y \in Y$. *Moreover,* $f_*\omega_{X/Y}^{\otimes m}$
is locally free for every nonnegative integer m.

Theorem 2.3.55 was first obtained in [Si]. For a simplified proof, see [Păŭ].
Theorem 2.3.55 is a clever application of the Ohsawa–Takegoshi L^2-extension theorem.

2.3.8 Related Conjectures

Let us recall the definition of pseudo-effective divisors in order to explain some conjectures on the minimal model program for higher-dimensional algebraic varieties.

Definition 2.3.56 (*Pseudo-effective divisors*) Let D be a Cartier divisor on a smooth
projective variety X. Then D is *pseudo-effective* if $D + A$ is big for any ample $\mathbb{Q}$-
divisor A on X.

The following conjecture is one of the most important conjectures in the theory
of minimal models for higher-dimensional algebraic varieties.

Conjecture 2.3.57 (*Nonvanishing conjecture*) Let X be a smooth projective variety
such that K_X is pseudo-effective. Then $\kappa(X) \geq 0$.

Once we know that Conjecture 2.3.57 holds true in dimension $\leq n$, then we can
freely run the minimal model program with ample scaling in dimension $\leq n$ (see
[Bi1] and [Has1]).

Let us recall Nakayama's numerical dimension κ_σ for the reader's convenience.
Although we do not need κ_σ in this book, we will use it in order to explain Conjecture
2.3.59 below and some recent developments in Sect. 5.3.

Definition 2.3.58 (*Nakayama's numerical dimension, see* [N4, Chap. V, 2.5.
Definition]) Let D be a pseudo-effective divisor on a smooth projective variety
X and let A be a Cartier divisor on X. If $H^0(X, \mathcal{O}_X(mD + A)) \neq 0$ for infinitely
many positive integers m, then we set

$$\sigma(D; A) = \max\left\{k \in \mathbb{Z}_{\geq 0} \;\middle|\; \limsup_{m \to \infty} \frac{\dim_{\mathbb{C}} H^0(X, \mathcal{O}_X(mD + A))}{m^k} > 0\right\}.$$

If $H^0(X, \mathcal{O}_X(mD + A)) \neq 0$ only for finitely many positive integers m, then we set $\sigma(D; A) = -\infty$. We define *Nakayama's numerical dimension* κ_σ by

$$\kappa_\sigma(X, D) = \max\{\sigma(D; A) \mid A \text{ is a Cartier divisor on } X\}.$$

It is well known that $\kappa_\sigma(X, D) \geq 0$ (see, for example, [N4, Chap. V, 2.7. Proposition]). If D is not pseudo-effective, then we put $\kappa_\sigma(X, D) = -\infty$. By this convention, we can define $\kappa_\sigma(X, D)$ for any Cartier divisor D on X. By definition, it is obvious that

$$\kappa(X, D) \leq \kappa_\sigma(X, D)$$

always holds for every Cartier divisor D on a smooth projective variety X. We say that a pseudo-effective divisor D is *abundant* when the equality $\kappa(X, D) = \kappa_\sigma(X, D)$ holds.

The following conjecture is a special case of the generalized abundance conjecture for projective lc pairs.

Conjecture 2.3.59 (*Generalized abundance conjecture*) Let X be a smooth projective variety and let Δ be a simple normal crossing divisor on X. If $K_X + \Delta$ is pseudo-effective, then $K_X + \Delta$ is abundant, that is,

$$\kappa(X, K_X + \Delta) = \kappa_\sigma(X, K_X + \Delta)$$

holds. Note that κ_σ denotes Nakayama's numerical dimension in Definition 2.3.58.

We note that Conjecture 2.3.59 contains Conjecture 2.3.57 as a special case.

Here, we do not discuss the precise definition of good minimal models or the details of the good minimal model conjecture. We just explain a special case of the good minimal model conjecture.

Conjecture 2.3.60 (*Good minimal model conjecture*) Let X be a smooth projective variety and let Δ be a simple normal crossing divisor on X. Assume that $K_X + \Delta$ is pseudo-effective. Then (X, Δ) has a good minimal model.

For the precise definition of good minimal models, see, for example, [F12, Definition 4.8.6]. By definition and some basic properties of κ and κ_σ, we have:

Proposition 2.3.61 *Let X be a smooth projective variety and let Δ be a simple normal crossing divisor on X. If (X, Δ) has a good minimal model, then $K_X + \Delta$ is abundant, that is, $\kappa_\sigma(X, K_X + \Delta) = \kappa(X, K_X + \Delta)$ holds.*

Proposition 2.3.61 says that Conjecture 2.3.59 follows from Conjecture 2.3.60. For the details of the generalized abundance conjecture and the good minimal model conjecture, see [F12, Sects. 4.8 and 4.11].

Precisely speaking, from the minimal model theoretic viewpoint, we have to formulate Conjecture 2.3.59 under the weaker assumption that Δ is an $\mathbb{R}$-divisor such that (X, Δ) is lc (see [F12, Conjecture 4.11.1]). When Δ is an $\mathbb{R}$-divisor, we have to use the *invariant Iitaka dimension* κ_ι (see [F12, Definition 2.5.5]) in order to formulate Conjecture 2.3.59. We recommend the reader to see [F12, Sects. 2.5 and 4.11] for the details. We also recommend the interested reader to see [HH] for some recent developments of the minimal model program for lc pairs.

Chapter 3
Viehweg's Weak Positivity

In this chapter, we describe the basic properties of Viehweg's weakly positive sheaves and big sheaves. In Sect. 3.1, we recall nef locally free sheaves, weakly positive sheaves and big sheaves and explain their basic properties. We also explain Viehweg's clever base change arguments. They are useful for various geometric applications. In Sect. 3.2, we prove a special case of the effective freeness of Popa–Schnell, which is relatively new. Although it can be proved as an easy application of (a generalization of) Kollár's vanishing theorem, it is very powerful. It greatly simplifies some parts of Viehweg's arguments and is indispensable in Sect. 4.3 in Chap. 4. Section 3.3 is devoted to Viehweg's results on direct images of relative pluricanonical bundles and adjoint bundles. In Sect. 3.4, we quickly explain some further developments and related topics without proof.

3.1 Weakly Positive Sheaves and Big Sheaves

In this section, we discuss the basic properties of weakly positive sheaves and big sheaves. We also discuss Viehweg's base change trick. Almost everything is contained in Viehweg's papers [Vi3], and [Vi4].

Definition 3.1.1 (*Weak positivity and bigness*) Let $\mathcal{F}$ be a torsion-free coherent sheaf on a smooth quasi-projective variety W. We say that $\mathcal{F}$ is *weakly positive* if, for every positive integer α and every ample invertible sheaf $\mathcal{H}$, there exists a positive integer β such that $\widehat{S}^{\alpha\beta}(\mathcal{F}) \otimes \mathcal{H}^{\otimes\beta}$ is generically generated by global sections. We say that a nonzero torsion-free coherent sheaf $\mathcal{F}$ is *big* (in the sense of Viehweg) if, for every ample invertible sheaf $\mathcal{H}$, there exists a positive integer a such that $\widehat{S}^{a}(\mathcal{F}) \otimes \mathcal{H}^{\otimes-1}$ is weakly positive.

O. Fujino, *Iitaka Conjecture*, SpringerBriefs in Mathematics,
https://doi.org/10.1007/978-981-15-3347-1_3

Note that there are several different definitions of weak positivity (see [Mo2, (5.1) Definition], Definitions 3.4.1, 3.4.8, and so on).

Remark 3.1.2 If $\widehat{S}^{\alpha\beta}(\mathcal{F}) \otimes \mathcal{H}^{\otimes\beta}$ is generically generated by global sections, then $\widehat{S}^{\alpha\beta\gamma}(\mathcal{F}) \otimes \mathcal{H}^{\otimes\beta\gamma}$ is also generically generated by global sections for every positive integer γ.

Although Remark 3.1.2 seems to be sufficient for almost all applications, we note the following lemma.

Lemma 3.1.3 ([Vi4, Lemma 3.2. (ii)]) *Let $\mathcal{F}$ be a weakly positive sheaf on a smooth quasi-projective variety W. Let α be a positive integer and let $\mathcal{H}$ be an ample invertible sheaf on W. Then there exists a positive integer β_0 such that for every positive integer $\beta > \beta_0$ the sheaf $\widehat{S}^{\alpha\beta}(\mathcal{F}) \otimes \mathcal{H}^{\otimes\beta}$ is generically generated by global sections.*

Proof By definition, there is a positive integer β' such that $\widehat{S}^{2\alpha\beta'\gamma}(\mathcal{F}) \otimes \mathcal{H}^{\otimes\beta'\gamma}$ is generically generated by global sections for every positive integer γ (see Remark 3.1.2). Since $\mathcal{H}$ is ample, we can find a positive integer c such that $\widehat{S}^{\alpha l}(\mathcal{F}) \otimes \mathcal{H}^{\otimes\beta'\gamma+l}$ is generated by global sections for every $1 \le l \le 2\beta'$ and for $\gamma \ge c$. Therefore, $\widehat{S}^{\alpha(2\beta'\gamma+l)}(\mathcal{F}) \otimes \mathcal{H}^{\otimes 2\beta'\gamma+l}$ is generically generated by global sections for every $\gamma \ge c$. Thus we can take $\beta_0 = 2c\beta'$, that is, $\widehat{S}^{\alpha\beta}(\mathcal{F}) \otimes \mathcal{H}^{\otimes\beta}$ is generically generated by global sections for every $\beta > \beta_0$. $\qquad\square$

Remark 3.1.4 Let $\mathcal{L}$ be an invertible sheaf on a smooth projective variety X. Then $\mathcal{L}$ is weakly positive if and only if $\mathcal{L}$ is pseudo-effective. We also note that $\mathcal{L}$ is big in the sense of Definition 3.1.1 if and only if $\mathcal{L}$ is big in the usual sense, that is, $\kappa(X, \mathcal{L}) = \dim X$.

The following lemmas are almost obvious by definition. We will use them repeatedly without mentioning them explicitly.

Lemma 3.1.5 *Let $\mathcal{F}$ be a torsion-free coherent sheaf on a smooth quasi-projective variety W and let $\mathcal{A}$ be a semi-ample invertible sheaf on W. Assume that $\mathcal{F}$ is weakly positive. Then $\mathcal{F} \otimes \mathcal{A}$ is weakly positive.*

Proof Let α be any positive integer and let $\mathcal{H}$ be any ample invertible sheaf on W. Then there exists a positive integer β such that $\widehat{S}^{\alpha\beta}(\mathcal{F}) \otimes \mathcal{H}^{\otimes\beta}$ is generically generated by global sections. By Remark 3.1.2, we may assume that $\mathcal{A}^{\otimes\beta}$ is generated by global sections after replacing β with a multiple. Therefore,

$$\widehat{S}^{\alpha\beta}(\mathcal{F} \otimes \mathcal{A}) \otimes \mathcal{H}^{\otimes\beta} = \widehat{S}^{\alpha\beta}(\mathcal{F}) \otimes \mathcal{H}^{\otimes\beta} \otimes \mathcal{A}^{\otimes\alpha\beta}$$

is generically generated by global sections. This means that the torsion-free coherent sheaf $\mathcal{F} \otimes \mathcal{A}$ is weakly positive. $\qquad\square$

Lemma 3.1.6 *Let $\mathcal{F}$ be a torsion-free coherent sheaf on a smooth quasi-projective variety W and let $\mathcal{B}$ be an ample invertible sheaf on W. Then $\mathcal{F} \otimes \mathcal{B}^{\otimes k}$ is weakly positive for some positive integer k.*

Proof Since $\mathcal{B}$ is ample, there exists a positive integer k such that $\mathcal{F} \otimes \mathcal{B}^{\otimes k}$ is generated by global sections. Then we can easily see that $\mathcal{F} \otimes \mathcal{B}^{\otimes k}$ is weakly positive. □

We will use the notion of nef locally free sheaves in Sect. 4.3 in Chap. 4.

Definition 3.1.7 (*Nefness*) Let $\mathcal{L}$ be an invertible sheaf on a smooth projective variety X. If $\mathcal{L} \cdot C \geq 0$ for every curve C on X, then $\mathcal{L}$ is said to be *nef*. Let $\mathcal{E}$ be a locally free sheaf of finite rank on a smooth projective variety X. If $\mathcal{E} = 0$ or $\mathcal{O}_{\mathbb{P}_X(\mathcal{E})}(1)$ is nef on $\mathbb{P}_X(\mathcal{E})$, then $\mathcal{E}$ is said to be *nef*.

Remark 3.1.8 A nef locally free sheaf was originally called a (*numerically*) *semipositive* locally free sheaf in the literature.

Lemma 3.1.9 *Let $\mathcal{E}$ be a locally free sheaf of finite rank on a smooth projective variety X. Then the following two conditions are equivalent:*

(i) *$\mathcal{E}$ is nef.*
(ii) *For every map $f : C \to X$ from a smooth projective curve C, every quotient invertible sheaf of $f^*\mathcal{E}$ has nonnegative degree.*

Proof The proof is almost obvious by the definition of $\mathbb{P}_X(\mathcal{E})$ and $\mathcal{O}_{\mathbb{P}_X(\mathcal{E})}(1)$. □

We collect some basic properties of nef locally free sheaves for the reader's convenience. For the details of nef locally free sheaves, see [Vi7, Sect. 2.2] and [Laz2, Chap. 6].

Lemma 3.1.10 *Let $\mathcal{E}$, $\mathcal{F}$, and $\mathcal{G}$ be locally free sheaves of finite rank on a smooth projective variety X. Then we have the following properties:*

(i) *Let $\mathcal{F} \to \mathcal{G}$ be a surjective morphism. If $\mathcal{F}$ is nef, then $\mathcal{G}$ is also nef.*
(ii) *Let*

$$0 \longrightarrow \mathcal{E} \longrightarrow \mathcal{F} \longrightarrow \mathcal{G} \longrightarrow 0$$

be a short exact sequence. If $\mathcal{E}$ and $\mathcal{G}$ are nef, then $\mathcal{F}$ is also nef.
(iii) *Let $\tau : X' \to X$ be a morphism from a smooth projective variety X'. If $\mathcal{E}$ is nef, then $\tau^*\mathcal{E}$ is also nef.*
(iv) *$\mathcal{E}$ is nef if and only if $f^*\mathcal{E} \otimes \mathcal{H}$ is ample for every map $f : C \to X$ from a smooth projective curve C and for every ample invertible sheaf $\mathcal{H}$ on C.*
(v) *If $\mathcal{E}$ is nef, then $\det \mathcal{E}$ and $S^a(\mathcal{E})$ are nef for every positive integer a.*
(vi) *If $\mathcal{E}$ and $\mathcal{F}$ are nef, then $\mathcal{E} \otimes \mathcal{F}$ is nef.*

Proof By (ii) in Lemma 3.1.9, we see that (i) and (iii) are almost obvious.

Let us prove (ii). Let $f : C \to X$ be a map from a smooth projective curve C and let $\mathcal{L}$ be a quotient invertible sheaf of $f^*\mathcal{F}$. If the composition map $f^*\mathcal{E} \to f^*\mathcal{F} \to \mathcal{L}$ is not zero, then the degree of $\mathcal{L}$ is nonnegative by the nefness of $\mathcal{E}$. If the map $f^*\mathcal{E} \to \mathcal{L}$ is zero, then $\mathcal{L}$ can be seen as a quotient invertible sheaf of $f^*\mathcal{G}$. Then the degree of $\mathcal{L}$ is nonnegative by the nefness of $\mathcal{G}$. Thus, the degree of $\mathcal{L}$ is always nonnegative. Therefore, $\mathcal{F}$ is nef by Lemma 3.1.9. Hence we get (ii).

From now on, we prove (iv). If $\mathcal{E}$ is nef, then $f^*\mathcal{E}$ is also nef by (iii). Therefore, $f^*\mathcal{E} \otimes \mathcal{H}$ is ample for every ample invertible sheaf $\mathcal{H}$ on C. Let $\mathcal{L}$ be a quotient invertible sheaf of $f^*\mathcal{E}$. We take an ample invertible sheaf $\mathcal{H}$ on X with $\deg \mathcal{H} = 1$. Then $\mathcal{L} \otimes \mathcal{H}$ is a quotient invertible sheaf of $f^*\mathcal{E} \otimes \mathcal{H}$ and is ample by the ampleness of $f^*\mathcal{E} \otimes \mathcal{H}$. Therefore, the degree of $\mathcal{L} \otimes \mathcal{H}$ is positive. Thus, the degree of $\mathcal{L}$ is at least $1 - \deg \mathcal{H} = 0$. This means that $\mathcal{E}$ is nef by Lemma 3.1.9.

Let us prove (v). Let $f : C \to X$ be a map from a smooth projective curve C and let $\mathcal{H}$ be an ample invertible sheaf on C. We take a finite map $g : C' \to C$ from a smooth projective curve C' such that $g^*\mathcal{H} \simeq \mathcal{H}'^{\otimes a}$ for some ample invertible sheaf $\mathcal{H}'$ on C' (see Lemma 2.3.38 and Lemma 3.3.6 below). Then we have

$$g^* \left(f^* S^a(\mathcal{E}) \otimes \mathcal{H} \right) \simeq S^a(g^* f^*\mathcal{E}) \otimes \mathcal{H}'^{\otimes a} \simeq S^a \left(g^* f^*\mathcal{E} \otimes \mathcal{H}' \right).$$

By (iv), $g^* f^*\mathcal{E} \otimes \mathcal{H}'$ is ample. By [Har1, Corollary (5.3)] (see also [Laz2, Theorem 6.1.15]), $S^a \left(g^* f^*\mathcal{E} \otimes \mathcal{H}' \right) \simeq g^* (f^* S^a(\mathcal{E}) \otimes \mathcal{H})$ is also ample. Since g is finite, we obtain that $f^* S^a(\mathcal{E}) \otimes \mathcal{H}$ is ample. By (iv), we see that $S^a(\mathcal{E})$ is nef. Since $\det \mathcal{E}$ is a quotient invertible sheaf of $S^r(\mathcal{E})$, where $r = \operatorname{rank}\mathcal{E}$, $\det \mathcal{E}$ is nef by (i).

Finally, we prove (vi). If $\mathcal{E}$ and $\mathcal{F}$ are nef, then $\mathcal{E} \oplus \mathcal{F}$ is nef by (ii). By (v), $S^2(\mathcal{E} \oplus \mathcal{F})$ is also nef. Since $\mathcal{E} \otimes \mathcal{F}$ is a direct summand of $S^2(\mathcal{E} \oplus \mathcal{F})$, $\mathcal{E} \otimes \mathcal{F}$ is nef by (i). $\qquad\qquad\square$

Remark 3.1.11 Let $\mathcal{E}$ be a nef locally free sheaf on a smooth projective variety X. Let $\mathcal{H}$ be an ample invertible sheaf on X and let α be a positive integer. Note that $\mathcal{O}_{\mathbb{P}_X(\mathcal{E})}(\alpha) \otimes \pi^*\mathcal{H}$ is an ample invertible sheaf on $\mathbb{P}_X(\mathcal{E})$, where $\pi : \mathbb{P}_X(\mathcal{E}) \to X$. We also note that $S^\alpha(\mathcal{E}) \otimes \mathcal{H}$ is ample. Therefore, there exists a positive integer β_0 such that $S^{\alpha\beta}(\mathcal{E}) \otimes \mathcal{H}^{\otimes \beta}$ is generated by global sections for every integer $\beta \geq \beta_0$. In particular, $\mathcal{E}$ is weakly positive.

We can easily check the following properties of weakly positive sheaves.

Lemma 3.1.12 ([Vi3, (1.3) Remark and Lemma 1.4]) *Let $\mathcal{F}$ and $\mathcal{G}$ be torsion-free coherent sheaves on a smooth quasi-projective variety W. Then we have the following properties:*

(i) *In order to check whether $\mathcal{F}$ is weakly positive, we may replace W with $W \setminus \Sigma$ for some closed subset Σ of codimension ≥ 2.*

(ii) *Let $\mathcal{F} \to \mathcal{G}$ be a generically surjective morphism. If $\mathcal{F}$ is weakly positive, then $\mathcal{G}$ is also weakly positive.*

(iii) *If $\widehat{S}^a(\mathcal{F})$ is weakly positive for some positive integer a, then $\mathcal{F}$ is weakly positive.*

(iv) *Let $\delta : W \to W''$ be a projective birational morphism to a smooth quasi-projective variety W'' and let E be a δ-exceptional Cartier divisor on W. If $\mathcal{F} \otimes \mathcal{O}_W(E)$ is weakly positive, then $\delta_*\mathcal{F}$ is weakly positive.*

(v) *Let $\tau : W' \to W$ be a finite surjective morphism from a smooth quasi-projective variety W'. If $\tau^*\mathcal{F}$ is weakly positive, then $\mathcal{F}$ is weakly positive.*

(vi) If $\mathcal{F}$ is weakly positive, then $\widehat{\det}(\mathcal{F})$ is weakly positive.
(vii) If $\mathcal{F}$ and $\mathcal{G}$ are weakly positive, then $\mathcal{F} \otimes \mathcal{G}/$torsion is also weakly positive.

Proof (i) and (ii) are obvious by the definition of weakly positive sheaves. Let us start the proof of (iii). The natural map

$$\widehat{S}^k \widehat{S}^l(\mathcal{F}) \to \widehat{S}^{kl}(\mathcal{F}) \tag{3.1.1}$$

is generically surjective, where k and l are any positive integers. Let α be a positive integer and let $\mathcal{H}$ be an ample invertible sheaf on W. Then there exists a positive integer β such that $\widehat{S}^{\alpha\beta}(\widehat{S}^a(\mathcal{F})) \otimes \mathcal{H}^{\otimes\beta}$ is generically generated by global sections if $\widehat{S}^a(\mathcal{F})$ is weakly positive. By using Remark 3.1.2, that is, by replacing β with a multiple, we may assume that $\mathcal{H}^{\otimes\beta(a-1)}$ is generated by global sections. Therefore, $\widehat{S}^{\alpha\beta}(\widehat{S}^a(\mathcal{F})) \otimes \mathcal{H}^{\otimes\beta a}$ is generically generated by global sections. By using the above generically surjective map (3.1.1), $\widehat{S}^{\alpha(a\beta)}(\mathcal{F}) \otimes \mathcal{H}^{\otimes(a\beta)}$ is generically generated by global sections. This means that $\mathcal{F}$ is weakly positive. Thus we get (iii). Let us prove (iv). Let $\mathcal{H}''$ be an ample invertible sheaf on W'' and let $\mathcal{H}$ be an ample invertible sheaf on W. We take a positive integer k such that

$$H^0(W, \delta^*\mathcal{H}''^{\otimes k} \otimes \mathcal{H}^{\otimes -1}) = H^0(W'', \mathcal{H}''^{\otimes k} \otimes \delta_*\mathcal{H}^{\otimes -1}) \neq 0.$$

For every positive integer α, $\widehat{S}^{\alpha k\beta}(\mathcal{F} \otimes \mathcal{O}_W(E)) \otimes \mathcal{H}^{\otimes\beta}$ is generically generated by global sections for some positive integer β since $\mathcal{F} \otimes \mathcal{O}_W(E)$ is weakly positive. Therefore, $\widehat{S}^{\alpha k\beta}(\mathcal{F} \otimes \mathcal{O}_W(E)) \otimes \delta^*\mathcal{H}''^{\otimes k\beta}$ is generically generated by global sections. Thus, we obtain that $\widehat{S}^{\alpha k\beta}(\delta_*\mathcal{F}) \otimes \mathcal{H}''^{\otimes k\beta}$ is generically generated by global sections. This implies that $\delta_*\mathcal{F}$ is weakly positive. This is (iv). Let $\mathcal{H}$ be an ample invertible sheaf on W. In order to prove (v), we may shrink W and may assume that $\mathcal{F}$ is locally free by (i). Since $\tau^*\mathcal{F}$ is weakly positive, we see that $S^{2\alpha\beta}(\tau^*\mathcal{F}) \otimes \tau^*\mathcal{H}^{\otimes\beta}$ is generically generated by global sections for every positive integer α and some large positive integer β. We note that we have a surjection

$$\tau_*\tau^*S^{2\alpha\beta}(\mathcal{F}) \otimes \mathcal{H}^{\otimes\beta} \to S^{2\alpha\beta}(\mathcal{F}) \otimes \mathcal{H}^{\otimes\beta}.$$

Hence we obtain a generically surjective morphism

$$\bigoplus_{\text{finite}} \tau_*\mathcal{O}_{W'} \otimes \mathcal{H}^{\otimes\beta} \to S^{2\alpha\beta}(\mathcal{F}) \otimes \mathcal{H}^{2\beta}.$$

We may assume that $\tau_*\mathcal{O}_{W'} \otimes \mathcal{H}^{\otimes\beta}$ is generated by global sections after replacing β with a multiple (see Remark 3.1.2). Thus $S^{2\alpha\beta}(\mathcal{F}) \otimes \mathcal{H}^{\otimes 2\beta}$ is generically generated by global sections. This means that $\mathcal{F}$ is weakly positive. So we obtain (v). Let $\mathcal{F}$ be a weakly positive sheaf. We put $r = \text{rank}(\mathcal{F})$. Let α be a positive integer and let $\mathcal{H}$ be an ample invertible sheaf. Then there exists a positive integer β such that $\widehat{S}^{\alpha\beta r}(\mathcal{F}) \otimes \mathcal{H}^{\otimes\beta}$ is generically generated. Hence $\widehat{\det}(\mathcal{F})^{\otimes\alpha b} \otimes \mathcal{H}^{\otimes b}$ is generically generated for $b = \text{rank}(\widehat{S}^{\alpha\beta r}(\mathcal{F}))\beta$. Thus, we obtain (vi). Since we do not use (vii) in this book, we omit the proof of (vii) here. For the proof, see [Vi4, Lemma 3.2. (iii)]. Note that the proof of (vii) is much harder than the proof of the other properties. □

We state the following theorem for the interested reader without proof. Note that we do not need Theorem 3.1.13 in this book.

Theorem 3.1.13 ([Vi4, Lemma 3.2. iii] and [Vi7, Corollary 2.20]) *Let $\mathcal{F}$ be a torsion-free coherent sheaf on a smooth quasi-projective variety W. Let T be a positive representation of $\mathrm{GL}(r, \mathbb{C})$ in the sense of Hartshorne (see [Har1, Sect. 5 Definition]), where $r = \mathrm{rank}\mathcal{F}$. We take a Zariski open set U of W such that $\mathcal{F}|_U$ is locally free and that $\mathrm{codim}_W(W \setminus U) \geq 2$ holds and put*

$$T(\mathcal{F}) := j_* T(\mathcal{F}|_U),$$

where $j : U \hookrightarrow W$ is the natural open immersion and $T(\mathcal{F}|_U)$ is a tensor bundle of $\mathcal{F}|_U$ associated to T (see [Har1, Sect. 5]). If $\mathcal{F}$ is weakly positive, then $T(\mathcal{F})$ is also weakly positive. In particular, $\widehat{S}^a(\mathcal{F})$ is weakly positive for every positive integer a when $\mathcal{F}$ is weakly positive.

By considering $\widehat{S}^2(\mathcal{F} \oplus \mathcal{G})$, we can obtain Lemma 3.1.12 (vii) as a corollary of Theorem 3.1.13. We omit the details of Theorem 3.1.13 here since we do not need this theorem in this book.

Remark 3.1.14 If $\mathcal{E}$ is a nef locally free sheaf on a smooth projective variety X, then $T(\mathcal{E})$ is also a nef locally free sheaf on X for any positive representation T of $\mathrm{GL}(r, \mathbb{C})$ with $r = \mathrm{rank}\mathcal{E}$, where $T(\mathcal{E})$ is a tensor bundle of $\mathcal{E}$ associated to T.

For bigness, we have the following lemma.

Lemma 3.1.15 ([Vi4, Lemma 3.6]) *Let $\mathcal{F}$ be a nonzero torsion-free coherent sheaf on a smooth quasi-projective variety W. Then the following three conditions are equivalent:*

(i) *There exist an ample invertible sheaf $\mathcal{H}$ on W, some positive integer v, and an inclusion $\bigoplus \mathcal{H} \hookrightarrow \widehat{S}^v(\mathcal{F})$, which is an isomorphism over a nonempty Zariski open set of W.*

(ii) *For every invertible sheaf $\mathcal{M}$ on W, there exists some positive integer γ such that $\widehat{S}^\gamma(\mathcal{F}) \otimes \mathcal{M}^{\otimes -1}$ is weakly positive. In particular, $\mathcal{F}$ is a big sheaf.*

(iii) *There exist some positive integer γ and an ample invertible sheaf $\mathcal{M}$ such that $\widehat{S}^\gamma(\mathcal{F}) \otimes \mathcal{M}^{\otimes -1}$ is weakly positive.*

Proof First, we assume (i). For every positive integer β, there exists a map $\bigoplus \mathcal{H}^{\otimes \beta} \to \widehat{S}^{\beta v}(\mathcal{F})$, which is generically surjective. If we choose β large enough, we may assume that $\mathcal{H}^{\otimes \beta} \otimes \mathcal{M}^{\otimes -1}$ is very ample. Therefore, $\widehat{S}^{\beta v}(\mathcal{F}) \otimes \mathcal{M}^{\otimes -1}$ is weakly positive by the generically surjective map $\bigoplus \mathcal{H}^{\otimes \beta} \otimes \mathcal{M}^{\otimes -1} \to \widehat{S}^{\beta v}(\mathcal{F}) \otimes \mathcal{M}^{\otimes -1}$ by Lemma 3.1.12 (ii). Thus we obtain (ii). Since (iii) is a special case of (ii), (iii) follows from (i).

Next, we assume (iii). If $\widehat{S}^\gamma(\mathcal{F}) \otimes \mathcal{M}^{\otimes -1}$ is weakly positive for some ample invertible sheaf $\mathcal{M}$ on W, then $\widehat{S}^{2\beta\gamma}(\mathcal{F}) \otimes \mathcal{M}^{\otimes -2\beta} \otimes \mathcal{M}^{\otimes \beta}$ is generically generated by global sections for some positive integer β. Thus we get a map

$$\bigoplus_{\text{finite}} \mathcal{M}^{\otimes \beta} \to \widehat{S}^{2\beta\gamma}(\mathcal{F}),$$

which is surjective over a nonempty Zariski open set of W. We choose $\mathrm{rank}\,(\widehat{S}^{2\beta\gamma}(\mathcal{F}))$ copies of $\mathcal{M}^{\otimes \beta}$ such that the corresponding sections generate the sheaf $\widehat{S}^{2\beta\gamma}(\mathcal{F}) \otimes \mathcal{M}^{\otimes -\beta}$ in the general point of W. Then we obtain (i) with $\mathcal{H} = \mathcal{M}^{\otimes \beta}$ and $\nu = 2\beta\gamma$. $\qquad\square$

Remark 3.1.16 First, we consider $\mathcal{E} = \mathcal{O}_{\mathbb{P}^1} \oplus \mathcal{O}_{\mathbb{P}^1}(1)$ and $X = \mathbb{P}_{\mathbb{P}^1}(\mathcal{E}) \to \mathbb{P}^1$. We put $\mathcal{O}_X(1) = \mathcal{O}_{\mathbb{P}_{\mathbb{P}^1}(\mathcal{E})}(1)$. Then

$$\dim H^0(X, \mathcal{O}_X(m)) = \dim H^0\left(\mathbb{P}^1, \bigoplus_{k=0}^m \mathcal{O}_{\mathbb{P}^1}(k)\right) = \frac{1}{2}(m+1)(m+2)$$

for every positive integer m. Therefore, $\mathcal{O}_X(1)$ is a big invertible sheaf on X. On the other hand, $\mathcal{E}$ is not big in the sense of Definition 3.1.1. This is because $S^m(\mathcal{E})$ contains $\mathcal{O}_{\mathbb{P}^1}$ as a direct summand for every positive integer m. Note that our definition of bigness is different from Lazarsfeld's (see [Laz2, Example 6.1.23]). Next, we put $\mathcal{F} = \mathcal{O}_{\mathbb{P}^1}(-1) \oplus \mathcal{O}_{\mathbb{P}^1}(1)$ and consider $Y = \mathbb{P}_{\mathbb{P}^1}(\mathcal{F}) \to \mathbb{P}^1$ with $\mathcal{O}_Y(1) = \mathcal{O}_{\mathbb{P}_{\mathbb{P}^1}(\mathcal{F})}(1)$. Then we can easily check that $\mathcal{O}_Y(1)$ is big as before. Of course, $\mathcal{O}_Y(1)$ is pseudo-effective. However, $S^{\alpha\beta}(\mathcal{F}) \otimes \mathcal{O}_{\mathbb{P}^1}(1)^{\otimes \beta}$ is not generically generated by global sections for $\alpha \geq 2$. Therefore, $\mathcal{F} = \mathcal{O}_{\mathbb{P}^1}(-1) \oplus \mathcal{O}_{\mathbb{P}^1}(1)$ is not weakly positive in the sense of Definition 3.1.1.

Remark 3.1.17 Let $\mathcal{E}$ be a nonzero locally free sheaf on a smooth projective variety X such that $\mathcal{E}$ is weakly positive. We consider $\pi : Y = \mathbb{P}_X(\mathcal{E}) \to X$ with $\mathcal{O}_Y(1) = \mathcal{O}_{\mathbb{P}_X(\mathcal{E})}(1)$. Then $\mathcal{O}_Y(1)$ is pseudo-effective. We can check this fact as follows. Let $\mathcal{H}$ be an ample invertible sheaf on X and let α be an arbitrary positive integer. Then we can take a positive integer β such that $S^{\alpha\beta}(\mathcal{E}) \otimes \mathcal{H}^{\otimes \beta}$ is generically generated by global sections since $\mathcal{E}$ is weakly positive. Thus, we have

$$H^0(Y, \mathcal{O}_Y(\alpha\beta) \otimes \pi^* \mathcal{H}^{\otimes \beta}) = H^0(X, S^{\alpha\beta}(\mathcal{E}) \otimes \mathcal{H}^{\otimes \beta}) \neq 0.$$

This implies that $\mathcal{O}_Y(1)$ is pseudo-effective by taking $\alpha \to \infty$.

3.1.18 (Viehweg's base change trick) Let us discuss Viehweg's clever base change arguments. They are very useful and important. The following results are contained in [Vi3, Sect. 3]. We closely follow [Mo2, Sect. 4].

Lemma 3.1.19 *Let V be a reduced Gorenstein scheme. Note that V may be reducible. We consider*

$$\rho : V' \xrightarrow{\delta} V^{\nu} \xrightarrow{\nu} V,$$

where $\nu : V^{\nu} \to V$ is the normalization and $\delta : V' \to V^{\nu}$ is a resolution of singularities. Then, for every positive integer n, we have

$$\nu_* \mathcal{O}_{V^{\nu}}(nK_{V^{\nu}}) \subset \omega_V^{\otimes n} \tag{3.1.2}$$

and

$$\delta_* \mathcal{O}_{V'}(nK_{V'} + E) \subset \mathcal{O}_{V^{\nu}}(nK_{V^{\nu}}), \tag{3.1.3}$$

where E is any δ-exceptional divisor on V'. In particular, we have

$$\rho_* \mathcal{O}_{V'}(nK_{V'} + E) \subset \omega_V^{\otimes n} \tag{3.1.4}$$

for every positive integer n. If U is a Zariski open set of V such that ρ is an isomorphism over U, then the inclusion (3.1.4) is an isomorphism over U.

Furthermore, if V has only rational singularities, then we have $\omega_V^{\otimes n} = \rho_* \omega_{V'}^{\otimes n}$ for every positive integer n.

Proof In Steps 1 and 2, we will prove (3.1.3) and (3.1.2), respectively.

Step 1 By taking the double dual of $\delta_* \mathcal{O}_{V'}(nK_{V'} + E)$, we obtain $\mathcal{O}_{V^{\nu}}(nK_{V^{\nu}})$. Therefore, we have

$$\delta_* \mathcal{O}_{V'}(nK_{V'} + E) \subset \mathcal{O}_{V^{\nu}}(nK_{V^{\nu}})$$

for every integer n.

Step 2 Since ν is birational, the trace map $\nu_* \mathcal{O}_{V^{\nu}}(K_{V^{\nu}}) \to \omega_V$ is a generically isomorphic injection

$$\nu_* \mathcal{O}_{V^{\nu}}(K_{V^{\nu}}) \hookrightarrow \omega_V. \tag{3.1.5}$$

Since ν is finite,

$$\nu^* \nu_* \mathcal{O}_{V^{\nu}}(K_{V^{\nu}}) \to \mathcal{O}_{V^{\nu}}(K_{V^{\nu}}) \tag{3.1.6}$$

is surjective and the kernel of (3.1.6) is the torsion part of $\nu^* \nu_* \mathcal{O}_{V^{\nu}}(K_{V^{\nu}})$. Therefore, by (3.1.5), we get an inclusion

$$\mathcal{O}_{V^{\nu}}(K_{V^{\nu}}) \hookrightarrow \nu^* \omega_V. \tag{3.1.7}$$

Let n be a positive integer with $n \geq 2$. Then we have

$$\mathcal{O}_{V^{\nu}}(nK_{V^{\nu}}) = \mathcal{O}_{V^{\nu}}(K_{V^{\nu}} + (n-1)K_{V^{\nu}}) \hookrightarrow \mathcal{O}_{V^{\nu}}(K_{V^{\nu}}) \otimes \nu^* \omega_V^{\otimes n-1}$$

by (3.1.7). Therefore, by taking ν_*, we get

$$\nu_* \mathcal{O}_{V^{\nu}}(nK_{V^{\nu}}) \hookrightarrow \nu_* \mathcal{O}_{V^{\nu}}(K_{V^{\nu}}) \otimes \omega_V^{\otimes n-1} \hookrightarrow \omega_V^{\otimes n}$$

by (3.1.5). This is what we wanted.

By the above construction of (3.1.2) and (3.1.3), it is obvious that the inclusion

$$\rho_* \mathcal{O}_{V'}(n K_{V'} + E) \subset \omega_V^{\otimes n}$$

is an isomorphism over U.

Step 3 We further assume that V has only rational singularities. Then it is well known that V has only canonical Gorenstein singularities (see Lemma 2.3.13). Therefore, we have $\omega_V^{\otimes n} = \rho_* \omega_{V'}^{\otimes n}$ for every positive integer n.

We complete the proof of Lemma 3.1.19. $\square$

Lemma 3.1.20 (Base Change Theorem, see [Mo2, (4.10)]) *Let $f : V \to W$ be a projective surjective morphism between smooth quasi-projective varieties. Let $\tau : W' \to W$ be a flat projective surjective morphism from a smooth quasi-projective variety W'. We consider the following commutative diagram:*

$$
\begin{array}{ccccc}
V & \xleftarrow{\ \widetilde{\rho}\ } & \widetilde{V} & \xleftarrow{\ \rho\ } & V' \\
{\scriptstyle f}\downarrow & & {\scriptstyle \widetilde{f}}\downarrow & \swarrow {\scriptstyle f'} & \\
W & \xleftarrow{\ \tau\ } & W' & &
\end{array}
$$

where $\widetilde{V} = V \times_W W'$ and $\rho : V' \to \widetilde{V}$ is a resolution. Then we have the following properties:

(i) *There is an inclusion*

$$f'_* \omega_{V'/W'}^{\otimes n} \subset \tau^*(f_* \omega_{V/W}^{\otimes n})$$

for every positive integer n.
Let P be a codimension one point of W'. Assume that $\widetilde{V}$ has only rational singularities over a neighborhood of P. Then we have

$$f'_* \omega_{V'/W'}^{\otimes n} = \widetilde{f}_* \omega_{\widetilde{V}/W'}^{\otimes n} = \tau^*(f_* \omega_{V/W}^{\otimes n})$$

at P.

(ii) *Let P be a codimension one point of W'. If $\tau(P)$ is a codimension one point of W and f is semistable in a neighborhood of $\tau(P)$, then $\widetilde{V}$ has only rational Gorenstein singularities over a neighborhood of P.*

(iii) *For every positive integer n, there is an inclusion*

$$\tau_* f'_* \omega_{V'/W}^{\otimes n} \subset (f_* \omega_{V/W}^{\otimes n} \otimes \tau_* \omega_{W'/W}^{\otimes n})^{**},$$

which is an isomorphism at codimension one point P of W if f or τ is semistable in a neighborhood of P.

Proof Since τ is flat, $\widetilde{V}$ is a reduced Gorenstein scheme and $\omega_{\widetilde{V}/W'} = \widetilde{\rho}^* \omega_{V/W}$ by the flat base change theorem [Ve, Theorem 2] (see also [Har2], [Co], and so on). Then we have that $\tau^* f_* \omega_{V/W}^{\otimes n} = \widetilde{f}_* \omega_{\widetilde{V}/W}^{\otimes n}$ for every positive integer n by the flat base change theorem (see [Har3, Chap. III, Proposition 9.3]). If $f : V \to W$ has connected fibers, then $\widetilde{V}$ is irreducible. In general, however, $\widetilde{V}$ may be reducible. By Lemma 3.1.19, we have $\rho_* \omega_{V'/W'}^{\otimes n} \subset \omega_{\widetilde{V}/W'}^{\otimes n}$. Therefore, we obtain $f'_* \omega_{V'/W'}^{\otimes n} \subset \tau^* f_* \omega_{V/W}^{\otimes n}$ for every positive integer n. The latter statement in (i) is obvious by the above argument and Lemma 3.1.19.

For (ii), it is sufficient to prove that $\widetilde{V}$ has only rational singularities over a neighborhood of P. By shrinking W around $\tau(P)$, we may assume that $\tau(P)$ is a smooth divisor on W and that $f : V \to W$ is (weakly) semistable. By shrinking W' around P, we may assume that $P = \tau^{-1}(\tau(P))$ and that P is a smooth divisor on W'. Then we obtain that $\widetilde{f} : \widetilde{V} \to W'$ is weakly semistable by Lemma 2.3.49 (see also [AbK, Lemma 6.2]). Thus, $\widetilde{V}$ has only rational Gorenstein singularities (see Lemma 2.3.43).

By (i), we have

$$
\begin{aligned}
\tau_* f'_* \omega_{V'/W}^{\otimes n} &= \tau_* f'_* (\omega_{V'/W'}^{\otimes n} \otimes f'^* \omega_{W'/W}^{\otimes n}) \\
&= \tau_* ((f'_* \omega_{V'/W'}^{\otimes n}) \otimes \omega_{W'/W}^{\otimes n}) \\
&\subset \tau_* ((\tau^* f_* \omega_{V/W}^{\otimes n}) \otimes \omega_{W'/W}^{\otimes n}) \\
&\subset (f_* \omega_{V/W}^{\otimes n} \otimes \tau_* \omega_{W'/W}^{\otimes n})^{**}
\end{aligned}
$$

by the projection formula. This is nothing but the inclusion in (iii). Without loss of generality, we may shrink W and assume that f is also flat for (iii). Since (iii) is symmetric with respect to f and τ, it is enough to check that the inclusion is an equality at P when f is semistable in a neighborhood of P. Then, by (i) and (ii), we have the equality at P. □

Remark 3.1.21 We can use Lemma 3.1.20 under the weaker assumption that V is a disjoint union of smooth quasi-projective varieties and every irreducible component of V is dominant onto W by f.

3.1.22 (Viehweg's fiber product trick) Let $f : V \to W$ be a projective surjective morphism between smooth quasi-projective varieties and let

$$
V^s = V \times_W V \times_W \cdots \times_W V
$$

be the s-fold fiber product. Let $V^{(s)}$ be an arbitrary resolution of the components of V^s dominating W and let $f^{(s)} : V^{(s)} \to W$ be the induced morphism. Note that $f_*^{(s)} \omega_{V^{(s)}/W}^{\otimes n}$ is independent of the choice of resolution $V^{(s)}$ for every positive integer n. If $f : V \to W$ has connected fibers, then $V^{(s)}$ is irreducible. In general, however, $V^{(s)}$ may be reducible, that is, $V^{(s)}$ is a disjoint union of smooth varieties.

Corollary 3.1.23 ([Vi3, Lemma 3.5] and [Mo2, (4.11) Corollary]) *Let $f : V \to W$ be a projective surjective morphism between smooth quasi-projective varieties. Let* *s and n be arbitrary positive integers. Then there exists a generically isomorphic injection*

$$a : (f_*^{(s)} \omega_{V^{(s)}/W}^{\otimes n})^{**} \hookrightarrow \left(\overset{s}{\bigotimes} f_* \omega_{V/W}^{\otimes n} \right)^{**}.$$

Let P be a codimension one point of W such that f is semistable in a neighborhood of P. Then a is an isomorphism at P.

Proof Since deleting closed subsets of W of codimension ≥ 2 does not change the double dual of torsion-free sheaves, we may assume that $f^{(i)}$ are flat for $i = 1, 2, \ldots, s$ and that $f_* \omega_{V/W}^{\otimes n}$ is locally free on W.

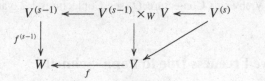

By Lemma 3.1.20 (iii), we obtain an injection

$$f_*^{(s)} \omega_{V^{(s)}/W}^{\otimes n} \hookrightarrow f_* \omega_{V/W}^{\otimes n} \otimes f_*^{(s-1)} \omega_{V^{(s-1)}/W}^{\otimes n}$$

such that the above injection is an isomorphism at P if f is semistable in a neighborhood of P. This proves the assertion by induction on s. □

3.1.24 Let $f : X \to Y$ be a surjective morphism between smooth projective varieties with connected fibers. Then we can always take a generically finite surjective morphism $\tau : Y' \to Y$ from a smooth projective variety Y' such that $f' : X' \to Y'$ is semistable in codimension one (see [KKMS] and [Vi3, Proposition 6.1]) or $f' : X' \to Y'$ factors through a weak semistable reduction $f^\dagger : X^\dagger \to Y'$ (see Theorem 2.3.46 and [AbK, Theorem 0.3]), where X' is a resolution of the main component of $X \times_Y Y'$.

$$\begin{array}{ccc} X & \longleftarrow & X' \\ f \downarrow & & \downarrow f' \\ Y & \longleftarrow{\scriptstyle \tau} & Y' \end{array}$$

Lemma 3.1.25 *In the notation in 3.1.24, if $f'_* \omega_{X'/Y'}^{\otimes n}$ is big for some positive integer n, then $f_* \omega_{X/Y}^{\otimes n}$ is also big.*

Proof Let $\mathcal{H}$ be an ample invertible sheaf on Y. Then there exists a positive integer a such that $\widehat{S}^a (f'_* \omega_{X'/Y'}^{\otimes n}) \otimes \tau^* \mathcal{H}^{\otimes -1}$ is weakly positive by Lemma 3.1.15. By removing a suitable closed subset Σ of codimension ≥ 2 from Y, we assume that $f_* \omega_{X/Y}^{\otimes n}$ is locally free and that τ is finite and flat. Then, by Lemma 3.1.20, we obtain a generically isomorphic injection

$$\widehat{S}^a(f'_*\omega_{X'/Y'}^{\otimes n}) \otimes \tau^*\mathcal{H}^{\otimes -1} \subset \tau^*(S^a(f_*\omega_{X/Y}^{\otimes n}) \otimes \mathcal{H}^{\otimes -1}).$$

By Lemma 3.1.12 (i), (ii), and (v), we see that $\widehat{S}^a(f_*\omega_{X/Y}^{\otimes n}) \otimes \mathcal{H}^{\otimes -1}$ is weakly positive. This means that $f_*\omega_{X/Y}^{\otimes n}$ is big.   □

We close this section with a useful observation.

3.1.26 By Lemma 3.1.25, we may assume that $f : X \to Y$ is semistable in codimension one or

$$f : X \xrightarrow{\ \delta\ } X^\dagger \xrightarrow{\ f^\dagger\ } Y$$

such that $f^\dagger : X^\dagger \to Y$ is weakly semistable and that δ is a resolution of singularities when we prove Viehweg's Conjecture Q (see Conjectures 1.2.6 and 4.1.2 below).

3.2   Effective Freeness Due to Popa–Schnell

In this section, we discuss the effective freeness due to Popa–Schnell (see [PopS]). The following statement is a special case of [PopS, Theorem 1.7]. It greatly simplifies some parts of Viehweg's arguments and will play a crucial role in Sect. 4.3 in Chap. 4 (see also [F10]).

Theorem 3.2.1 ([PopS, Theorem 1.4]) *Let* $f : X \to Y$ *be a surjective morphism from a smooth projective variety* X *to a projective variety* Y *with* $\dim Y = n$. *Let* k *be a positive integer and let* $\mathcal{L}$ *be an ample invertible sheaf on* Y *such that* $|\mathcal{L}|$ *is free. Then we have*

$$H^i(Y, f_*\omega_X^{\otimes k} \otimes \mathcal{L}^{\otimes l}) = 0$$

for every $i > 0$ *and every* $l \geq nk + k - n$. *By Castelnuovo–Mumford regularity,* $f_*\omega_X^{\otimes k} \otimes \mathcal{L}^{\otimes l}$ *is generated by global sections for every* $l \geq k(n+1)$.

The proof of Theorem 3.2.1 is surprisingly easy. Before we prove Theorem 3.2.1, we note the following remarks.

Remark 3.2.2 In Theorem 3.2.1, by Kollár's vanishing theorem (see [Kol1, Theorem 2.1 (iii)] and Theorem 2.3.50 (ii)), we have

$$H^i(Y, R^j f_*\omega_X \otimes \mathcal{A}) = 0$$

for every $i > 0$ and every j, where $\mathcal{A}$ is any ample invertible sheaf on Y. Therefore, by Castelnuovo–Mumford regularity (see Lemma 3.2.4 below), we obtain that $R^j f_*\omega_X \otimes \mathcal{L}^{\otimes l}$ is generated by global sections for every $l \geq n + 1$ and every j.

Remark 3.2.3 Takahiro Shibata proved that there are no positive integers N depending only on $\dim X$, $\dim Y$, $k \geq 2$, and $j \geq 1$ such that

$$H^i(Y, R^j f_* \omega_X^{\otimes k} \otimes \mathcal{L}^{\otimes l}) = 0$$

holds for every $i > 0$ and every $l \geq N$. For the details, see [Sh, Sect. 4]. This means that Theorem 3.2.1 does not hold for $R^j f_* \omega_X^{\otimes k}$ when $j \geq 1$ and $k \geq 2$.

Let us prove Theorem 3.2.1.

Proof of Theorem 3.2.1 Let us consider

$$\mathcal{M} = \mathrm{Im}\big(f^* f_* \omega_X^{\otimes k} \to \omega_X^{\otimes k}\big).$$

By taking blow-ups, we may assume that $\mathcal{M}$ is an invertible sheaf such that $\omega_X^{\otimes k} = \mathcal{M} \otimes \mathcal{O}_X(E)$ for some effective divisor E on X. We may further assume that $\mathrm{Supp}\, E$ is a simple normal crossing divisor. We can take the smallest integer $m \geq 0$ such that $f_* \omega_X^{\otimes k} \otimes \mathcal{L}^{\otimes m}$ is generated by global sections because $\mathcal{L}$ is ample. Then $\omega_X^{\otimes k} \otimes \mathcal{O}_X(-E) \otimes f^* \mathcal{L}^{\otimes m}$ is also generated by global sections. Note that $\omega_X^{\otimes k} \otimes \mathcal{O}_X(-E) = \mathcal{M}$, $f_* \mathcal{M} = f_* \omega_X^{\otimes k}$, and $f^* f_* \mathcal{M} \to \mathcal{M}$ is surjective. Therefore, we can take a smooth general effective divisor D such that $\mathrm{Supp}(D + E)$ is a simple normal crossing divisor on X and that

$$kK_X + mf^*L \sim D + E,$$

where $\mathcal{L} = \mathcal{O}_X(L)$. Thus we have

$$(k-1)K_X \sim_{\mathbb{Q}} \frac{k-1}{k}D + \frac{k-1}{k}E - \frac{k-1}{k}mf^*L.$$

So we obtain

$$kK_X - \left\lfloor \frac{k-1}{k}E \right\rfloor + lf^*L$$

$$\sim_{\mathbb{Q}} K_X + \frac{k-1}{k}D + \left\{ \frac{k-1}{k}E \right\} + \left(l - \frac{k-1}{k}m \right) f^*L.$$

By the vanishing theorem (see, for example, Theorem 2.3.50 (ii), [Kol5, 10.15 Corollary], [F7, Theorem 6.3 (ii)], and so on), which is a generalization of Kollár's vanishing theorem (see [Kol1, Theorem 2.1 (iii)]), we obtain

$$H^i(Y, f_* \omega_X^{\otimes k} \otimes \mathcal{L}^{\otimes l}) = 0$$

for every $i > 0$ if $l - \frac{k-1}{k}m > 0$. Note that

$$f_* \mathcal{O}_X \left(kK_X - \left\lfloor \frac{k-1}{k}E \right\rfloor \right) = f_* \omega_X^{\otimes k}$$

by the definition of E. Therefore, if $l > \frac{k-1}{k}m + n$, then $f_*\omega_X^{\otimes k} \otimes \mathcal{L}^{\otimes l}$ is generated by global sections by Castelnuovo–Mumford regularity (see Lemma 3.2.4 below). By the choice of m, we obtain

$$m \le \frac{k-1}{k}m + n + 1.$$

This implies $m \le k(n+1)$. Therefore, we obtain that if

$$l > \frac{k-1}{k} \cdot k(n+1) = kn + k - n - 1$$

then

$$H^i(Y, f_*\omega_X^{\otimes k} \otimes \mathcal{L}^{\otimes l}) = 0$$

for every $i > 0$. □

We recall the following lemma, which is an easy consequence of Castelnuovo–Mumford regularity, for the reader's convenience. We have already used it in the proof of Theorem 3.2.1.

Lemma 3.2.4 *Let V be a projective variety and let $\mathcal{A}$ be an ample invertible sheaf on V such that $|\mathcal{A}|$ is free. Let $\mathcal{F}$ be any coherent sheaf on V. If*

$$H^i(V, \mathcal{F} \otimes \mathcal{A}^{\otimes -i}) = 0$$

for every $i > 0$, then $\mathcal{F}$ is generated by global sections.

Proof This result is well known. For the details, see, for example, [Laz1, Theorem 1.8.5]. □

By combining Theorem 3.2.1 with Viehweg's fiber product trick (see Corollary 3.1.23), we can easily recover Viehweg's weak positivity theorem.

Theorem 3.2.5 (Viehweg's weak positivity theorem (see [Vi3, Theorem III])) *Let $f : X \to Y$ be a surjective morphism between smooth projective varieties. Then $f_*\omega_{X/Y}^{\otimes k}$ is weakly positive for every positive integer k.*

The following proof is due to Popa–Schnell (see [PopS]).

Proof Let $f^s : X^s = X \times_Y X \times_Y \cdots \times_Y X \to Y$ be the s-fold fiber product. Then we obtain a generically isomorphic injection

$$a : f_*^{(s)}\omega_{X^{(s)}/Y}^{\otimes k} \hookrightarrow \left(\bigotimes^s f_*\omega_{X/Y}^{\otimes k}\right)^{**}$$

for every $k \ge 1$ and every $s \ge 1$ by Corollary 3.1.23, where $X^{(s)} \to X^s$ is a resolution of the components of X^s dominating Y and $f^{(s)} : X^{(s)} \to Y$ is the induced morphism.

Note that $X^{(s)}$ may be a disjoint union of smooth projective varieties. Let $\mathcal{H}$ be any ample invertible sheaf on Y. We take a positive integer p such that $|\mathcal{H}^{\otimes p}|$ is basepoint-free. Then, by Theorem 3.2.1, we obtain that

$$f_*^{(s)} \omega_{X^{(s)}/Y}^{\otimes k} \otimes \omega_Y^{\otimes k} \otimes \mathcal{H}^{\otimes pk(n+1)}$$

is generated by global sections for every $s \geq 1$ and every $k \geq 1$, where $n = \dim Y$. From now on, we fix a positive integer k. We take a positive integer q such that $|\mathcal{H}^{\otimes r} \otimes \omega_Y^{\otimes -k}|$ is basepoint-free for every integer $r \geq q$. Then

$$\left(\bigotimes^{s} f_* \omega_{X/Y}^{\otimes k} \right)^{**} \otimes \mathcal{H}^{\otimes \beta}$$

is generically generated by global sections for every $\beta \geq q + pk(n+1)$ by the generically isomorphic injection a. Therefore, for every positive integer α,

$$\widehat{S}^{\alpha\beta}(f_* \omega_{X/Y}^{\otimes k}) \otimes \mathcal{H}^{\otimes \beta}$$

is generically generated by global sections for $\beta \geq q + pk(n+1)$. This implies that $f_* \omega_{X/Y}^{\otimes k}$ is weakly positive. $\qquad\square$

Remark 3.2.6 The proof of Theorem 3.2.5 says that

$$\left(\bigotimes^{s} f_* \omega_{X/Y}^{\otimes k} \right)^{**} \otimes \omega_Y^{\otimes k} \otimes \mathcal{A}^{\otimes k(n+1)}$$

is generated by global sections over U, where $\mathcal{A}$ is an ample invertible sheaf on Y such that $|\mathcal{A}|$ is free and U is a nonempty Zariski open set of Y such that f is smooth over U. Note that the inclusion a in the proof of Theorem 3.2.5 is an isomorphism over U. For a more general result, see Theorem 3.4.9 below, which is due to Dutta–Murayama (see [DM, Theorem E]). We also recommend the reader to see [F19, Theorems 1.5, 1.7, and 1.8] for some related results.

We close this section with an obvious corollary of Theorem 3.2.1.

Corollary 3.2.7 *Let $f : X \to Y$ be a surjective morphism from a projective variety X to a smooth projective variety Y. Assume that X has only rational Gorenstein singularities. Let $\mathcal{L}$ be an ample invertible sheaf on Y such that $|\mathcal{L}|$ is free and let k be a positive integer. Then*

$$f_* \omega_X^{\otimes k} \otimes \mathcal{L}^{\otimes l} = f_* \omega_{X/Y}^{\otimes k} \otimes \omega_Y^{\otimes k} \otimes \mathcal{L}^{\otimes l}$$

is generated by global sections for $l \geq k(\dim Y + 1)$.

Proof Since X has only rational Gorenstein singularities, X has only canonical Gorenstein singularities (see Lemma 2.3.13). Therefore, by replacing X with its resolution, we may assume that X is a smooth projective variety. Then this corollary follows from Theorem 3.2.1.   □

We will use this corollary in the proof of Theorem 4.3.1.

3.3 Direct Images of Relative Pluricanonical Bundles

In this section, let us discuss the weak positivity of direct images of adjoint bundles and relative (pluri-)canonical bundles, and some related topics. We closely follow [Vi3, Sect. 5] and [Vi4, Sect. 3].

Lemma 3.3.1 ([Vi3, Theorem 4.1]) *Let $f : V \to W$ be a surjective morphism between smooth projective varieties. Then $f_*\omega_{V/W}$ is weakly positive.*

This result is well known. We have already proved a more general result (see Theorem 3.2.5) by using the effective freeness due to Popa–Schnell (see Theorem 3.2.1), so we omit the detailed proof here. Note that this lemma can be proved without using the theory of variations of Hodge structure (see, for example, [Kol1] and [Vi5, 5. Weak positivity]). We can prove it as an application of Kollár's vanishing theorem (see also the proof of Theorem 3.2.5 and [F5, Sect. 5]).

For the reader's convenience, we give a sketch of the original proof of Lemma 3.3.1.

Sketch of the Proof of Lemma 3.3.1 Let Σ be a closed subset of W such that f is smooth over $W_0 = W \setminus \Sigma$. Let $\tau : W' \to W$ be a projective birational morphism from a smooth projective variety W' such that $\tau^{-1}(\Sigma)$ is a simple normal crossing divisor on W'. By Lemma 3.1.12 (iv), we can replace W with W'. In this situation, $f_*\omega_{V/W}$ is locally free and can be characterized as the upper canonical extension of a suitable Hodge bundle. By Lemma 3.1.12 (v), (ii), and the unipotent reduction theorem, we may further assume that all the local monodromies on $R^d f_{0*}\mathbb{C}_{V_0}$ around Σ are unipotent, where $d = \dim V - \dim W$ and $f_0 = f|_{V_0} : V_0 = f^{-1}(W_0) \to W_0$. In this case, we know that $f_*\omega_{V/W}$ is a nef locally free sheaf by the theory of variations of Hodge structure (see [Kaw3, Theorem 5]). Therefore, we obtain that $f_*\omega_{V/W}$ is weakly positive (see Remark 3.1.11).   □

Remark 3.3.2 The Hodge theoretic part of [Kaw3] seems to be insufficient. So we recommend the reader to see [FF1], [FFS], and [Fs] for the Hodge theoretic aspect of the nefness of $f_*\omega_{V/W}$ and some generalizations. We also recommend the reader to see [FF2], where we treat some analytic generalizations of the Fujita–Zucker–Kawamata semipositivity theorem.

The following lemma may look technical and artificial but is very important.

Lemma 3.3.3 ([Vi3, Lemma 5.1]) *Let $f : V \to W$ be a projective surjective morphism between smooth quasi-projective varieties. Let $\mathcal{L}$ and $\mathcal{M}$ be invertible sheaves* *on V and let E be an effective divisor on V such that $\operatorname{Supp} E$ is a simple normal crossing divisor. Assume that*

$$\mathcal{L}^{\otimes N} = \mathcal{M} \otimes \mathcal{O}_V(E)$$

for some positive integer N. We further assume that there exists a nonempty Zariski open set U of W such that some power of $\mathcal{M}$ is generated by global sections over $f^{-1}(U)$, that is,

$$H^0(V, \mathcal{M}^{\otimes k}) \otimes \mathcal{O}_V \to \mathcal{M}^{\otimes k}$$

is surjective on $f^{-1}(U)$ for some positive integer k. Then we obtain that

$$f_*(\omega_{V/W} \otimes \mathcal{L}^{(i)})$$

is weakly positive for $0 \le i \le N - 1$, where

$$\mathcal{L}^{(i)} = \mathcal{L}^{\otimes i} \otimes \mathcal{O}_V \left(-\left\lfloor \frac{iE}{N} \right\rfloor \right).$$

Proof Since the statement is compatible with replacing N by NN', E by $N'E$, and $\mathcal{M}$ by $\mathcal{M}^{\otimes N'}$ for some positive integer N', we may assume that $\mathcal{M}$ itself is generated by global sections over $f^{-1}(U)$. Without loss of generality, we may shrink U if necessary. Let $B + F$ be the zero set of a general section of $\mathcal{M}$ such that every irreducible component of B is dominant onto W and that $\operatorname{Supp} F \subset V \setminus f^{-1}(U)$. By Bertini's theorem, we may assume that B is smooth and $\operatorname{Supp}(B + E)$ is a simple normal crossing divisor on $f^{-1}(U)$. We note that $\mathcal{M} = \mathcal{O}_V(B + F)$. By taking a suitable birational modification outside $f^{-1}(U)$, we may assume that B is smooth and that $\operatorname{Supp}(B + E + F)$ is a simple normal crossing divisor (see, for example, [F12, Lemma 2.3.19]). In fact, if $\rho : V' \to V$ is a birational modification which is an isomorphism over $f^{-1}(U)$ and if $\mathcal{L}' = \rho^* \mathcal{L}$, $\mathcal{M}' = \rho^* \mathcal{M}$, and $E' = \rho^* E$, then we can easily check that $\rho_*(\omega_{V'} \otimes \mathcal{L}'^{(i)})$ is contained in $\omega_V \otimes \mathcal{L}^{(i)}$. By construction, $\rho_*(\omega_{V'} \otimes \mathcal{L}'^{(i)})$ coincides with $\omega_V \otimes \mathcal{L}^{(i)}$ on $f^{-1}(U)$. We note that $\mathcal{L}^N = \mathcal{O}_V(B + F + E)$ holds since $\mathcal{M} = \mathcal{O}_V(B + F)$. When we prove the weak positivity of $f_*(\omega_{V/W} \otimes \mathcal{L}^{(i)})$, by replacing E with $E + F$, we may assume that $F = 0$ (see Lemma 3.1.12 (ii)). Note that every irreducible component of F is vertical with respect to $f : V \to W$. We take a cyclic cover $p : Z' \to X$ associated to $\mathcal{L}^{\otimes N} = \mathcal{O}_V(B + E)$, that is, Z' is the normalization of $\operatorname{Spec}_X \bigoplus_{i=1}^{N-1} \mathcal{L}^{\otimes -i}$ (see 2.3.37). Let Z be a resolution of the cyclic cover Z' and let $g : Z \to W$ be the corresponding morphism.

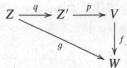

It is well known that Z' has only quotient singularities and

$$p_* q_* \omega_Z \simeq p_* \omega_{Z'} \simeq \bigoplus_{i=0}^{N-1} \omega_V \otimes \mathcal{L}^{(i)}.$$

Thus, we obtain

$$g_* \omega_{Z/W} \simeq \bigoplus_{i=0}^{N-1} f_*(\omega_{V/W} \otimes \mathcal{L}^{(i)}).$$

Therefore, by Lemmas 3.3.1 and 3.1.12 (ii), $f_*(\omega_{V/W} \otimes \mathcal{L}^{(i)})$ is weakly positive for every $0 \le i \le N - 1$. $\square$

As an application of Lemma 3.3.3, we have:

Lemma 3.3.4 ([Vi3, Corollary 5.2]) *Let $f : V \to W$ be a projective surjective morphism between smooth quasi-projective varieties and let $\mathcal{H}$ be an ample invertible sheaf on W such that $\widehat{S}^\nu(f_*\omega_{V/W}^{\otimes k} \otimes \mathcal{H}^{\otimes k})$ is generically generated by global sections for a given positive integer k and some positive integer ν. Then $f_*\omega_{V/W}^{\otimes k} \otimes \mathcal{H}^{\otimes k-1}$ is weakly positive.*

Proof If $f_*\omega_{V/W}^{\otimes k} = 0$, then the statement is obvious. Therefore, we may assume that $f_*\omega_{V/W}^{\otimes k} \ne 0$. By replacing W with $W \setminus \Sigma$, where Σ is a suitable closed subset of codimension ≥ 2, we may assume that f is flat and that $f_*\omega_{V/W}^{\otimes k}$ is locally free. We put $\mathcal{L} = \omega_{V/W} \otimes f^*\mathcal{H}$ and

$$\mathcal{M} = \mathrm{Im}\left(f^*(f_*\omega_{V/W}^{\otimes k} \otimes \mathcal{H}^{\otimes k}) \to \omega_{V/W}^{\otimes k} \otimes f^*\mathcal{H}^{\otimes k} \right).$$

By taking blow-ups, we may assume that $\mathcal{M}$ is invertible and that $\mathcal{L}^{\otimes k} = \mathcal{M} \otimes \mathcal{O}_V(E)$ for some effective divisor E on V such that $\mathrm{Supp}\, E$ is a simple normal crossing divisor. Since $\widehat{S}^\nu(f_*\omega_{V/W}^{\otimes k} \otimes \mathcal{H}^{\otimes k})$ is generically generated by global sections by assumption, we see that $\mathcal{M}^{\otimes \nu}$ is generated by global sections over $f^{-1}(U)$, where U is some nonempty Zariski open set of W. By Lemma 3.3.3, we obtain that $f_*(\omega_{V/W} \otimes \mathcal{L}^{(k-1)})$ is weakly positive, where

$$\mathcal{L}^{(k-1)} = \mathcal{L}^{\otimes k-1} \otimes \mathcal{O}_V\left(-\left\lfloor \frac{k-1}{k} E \right\rfloor \right).$$

Note that

$$\mathcal{M} \otimes f^*\mathcal{H}^{\otimes -1} \subset \omega_{V/W} \otimes \mathcal{L}^{(k-1)}$$

and that

$$f_*(\omega_{V/W} \otimes \mathcal{L}^{(k-1)}) \subset f_*\omega_{V/W}^{\otimes k} \otimes \mathcal{H}^{\otimes k-1}.$$

By the definition of $\mathcal{M}$, we have

$$f_* \mathcal{M} \otimes \mathcal{H}^{\otimes -1} = f_*(\omega_{V/W} \otimes \mathcal{L}^{(k-1)}) = f_* \omega_{V/W}^{\otimes k} \otimes \mathcal{H}^{\otimes k-1}.$$

Thus we obtain that

$$f_* \omega_{V/W}^{\otimes k} \otimes \mathcal{H}^{\otimes k-1}$$

is weakly positive.   □

By using Lemma 3.3.4, Viehweg cleverly obtained:

Theorem 3.3.5 (Viehweg's weak positivity theorem (see [Vi3, Theorem III])) *Let* $f : V \to W$ *be a surjective morphism between smooth projective varieties. Then* $f_* \omega_{V/W}^{\otimes k}$ *is weakly positive for every positive integer* k.

Note that we have already proved Theorem 3.3.5 by using the effective freeness due to Popa–Schnell (see Theorem 3.2.5). However, we give Viehweg's original proof here since it is interesting and useful for some other applications (see, for example, [F16]).

Proof We divide the proof into two steps.

Step 1 Let $\mathcal{H}$ be any ample invertible sheaf on W. We put

$$r = \min\{s \in \mathbb{Z}_{>0} \mid f_* \omega_{V/W}^{\otimes k} \otimes \mathcal{H}^{\otimes sk-1} \text{ is weakly positive}\}.$$

By definition, we can find a positive integer ν such that

$$\widehat{S}^\nu (f_* \omega_{V/W}^{\otimes k}) \otimes \mathcal{H}^{\otimes rk\nu - \nu} \otimes \mathcal{H}^{\otimes \nu}$$

is generically generated by global sections. By Lemma 3.3.4, we have that $f_* \omega_{V/W}^{\otimes k} \otimes \mathcal{H}^{\otimes rk-r}$ is weakly positive. If $r = 1$, then $k \geq r$ always holds since k is a positive integer. If $r \geq 2$, then the choice of r allows this only if $(r-1)k - 1 < rk - r$, equivalently, $r \leq k$. Thus, we always have $r \leq k$. Therefore, $f_* \omega_{V/W}^{\otimes k} \otimes \mathcal{H}^{\otimes k^2-k}$ is weakly positive.

Step 2 Let α be a positive integer. By Lemma 3.3.6 below, we can take a finite flat morphism $\tau : W' \to W$ from a smooth projective variety W' such that $\tau^* \mathcal{H} = \mathcal{H}'^{\otimes d}$ for $d = 2\alpha(k^2 - k) + 1$. We put $V' = V \times_W W'$. Then we may assume that V' is a smooth projective variety by Lemma 3.3.6 below. Let $f' : V' \to W'$ be the induced morphism.

$$\begin{array}{ccc} V & \longleftarrow & V' \\ {\scriptstyle f}\downarrow & & \downarrow{\scriptstyle f'} \\ W & \underset{\tau}{\longleftarrow} & W' \end{array}$$

By applying the result obtained in Step 1 to $f' : V' \to W'$, we obtain that

$$f'_* \omega^{\otimes k}_{V'/W'} \otimes \mathcal{H}'^{\otimes k^2 - k}$$

is weakly positive. Since $f'_* \omega^{\otimes k}_{V'/W'} = \tau^* f_* \omega^{\otimes k}_{V/W}$, we see that $\tau^* f_* \omega^{\otimes k}_{V/W} \otimes \mathcal{H}'^{\otimes k^2 - k}$ is weakly positive. Let β be a large positive integer such that

$$\widehat{S}^{2\alpha\beta}(\tau^* f_* \omega^{\otimes k}_{V/W} \otimes \mathcal{H}'^{\otimes k^2 - k}) \otimes \mathcal{H}'^{\otimes \beta} = \tau^* \widehat{S}^{2\alpha\beta}(f_* \omega^{\otimes k}_{V/W}) \otimes \tau^* \mathcal{H}^{\otimes \beta}$$

is generically generated by global sections. Let $\widehat{W}$ be a nonempty Zariski open set of W such that $\widehat{S}^{2\alpha\beta}(f_* \omega^{\otimes k}_{V/W})$ is locally free and that $\mathrm{codim}_W(W \setminus \widehat{W}) \geq 2$. By shrinking W, we may assume that $W = \widehat{W}$. Then we have a surjection

$$\tau_* \tau^* \widehat{S}^{2\alpha\beta}(f_* \omega^{\otimes k}_{V/W}) \otimes \mathcal{H}^{\otimes \beta} \to \widehat{S}^{2\alpha\beta}(f_* \omega^{\otimes k}_{V/W}) \otimes \mathcal{H}^{\otimes \beta}.$$

Therefore, we obtain a homomorphism

$$\tau_* \mathcal{O}_{W'} \otimes \mathcal{H}^{\otimes \beta} \to \widehat{S}^{2\alpha\beta}(f_* \omega^{\otimes k}_{V/W}) \otimes \mathcal{H}^{\otimes 2\beta}$$

which is surjective over a nonempty Zariski open set. Without loss of generality, we may assume that $\tau_* \mathcal{O}_{W'} \otimes \mathcal{H}^{\otimes \beta}$ is generated by global sections by replacing β with a multiple (see Remark 3.1.2). Thus, $\widehat{S}^{2\alpha\beta}(f_* \omega^{\otimes k}_{V/W}) \otimes \mathcal{H}^{\otimes 2\beta}$ is generated by global sections over a nonempty Zariski open set.

This means that $f_* \omega^{\otimes k}_{V/W}$ is weakly positive.   □

The following covering construction is very important and useful. We have already used it in the proof of Theorem 3.3.5. The description of Kawamata's covering trick in [EsV, 3.19. Lemma] is very useful for our purpose (see also Lemma 2.3.38, [AbK, 5.3. Kawamata's covering], and [Vi7, Lemma 2.5]).

Lemma 3.3.6 *Let $f : V \to W$ be a projective surjective morphism between smooth quasi-projective varieties and let H be a Cartier divisor on W. Let d be an arbitrary positive integer. Then we can take a finite flat morphism $\tau : W' \to W$ from a smooth quasi-projective variety W' and a Cartier divisor H' on W' such that $\tau^* H \sim dH'$ and that $V' = V \times_W W'$ is a smooth quasi-projective variety with $\omega_{V'/W'} = \rho^* \omega_{V/W}$, where $\rho : V' \to V$.*

Proof We take general very ample Cartier divisors D_1 and D_2 with the following properties:

  (i) $H \sim D_1 - D_2$,
  (ii) D_1, D_2, $f^* D_1$, and $f^* D_2$ are smooth,
  (iii) D_1 and D_2 have no common components, and
  (iv) $\mathrm{Supp}(D_1 + D_2)$ and $\mathrm{Supp}(f^* D_1 + f^* D_2)$ are simple normal crossing divisors.

We take a finite flat cover due to Kawamata with respect to W and $D_1 + D_2$ (see Lemma 2.3.38). Then we obtain $\tau : W' \to W$ and H' such that $\tau^* H \sim dH'$. By the construction of the above Kawamata cover $\tau : W' \to W$, we may assume that the ramification locus Σ of τ in W is a general simple normal crossing divisor. This means that f^*P is a smooth divisor for any irreducible component P of Σ and that $f^*\Sigma$ is a simple normal crossing divisor on V. In this situation, we can easily check that $V' = V \times_W W'$ is a smooth quasi-projective variety.

$$
\begin{array}{ccc}
V' & \xrightarrow{\ \rho\ } & V \\
\downarrow{\scriptstyle f'} & & \downarrow{\scriptstyle f} \\
W' & \xrightarrow{\ \tau\ } & W
\end{array}
$$

By construction, we can also easily check that $\omega_{V'/W'} = \rho^*\omega_{V/W}$ by the Hurwitz formula.

Let us see the construction of $f' : V' \to W'$ more precisely for the reader's convenience. Let $\mathcal{A}$ be an ample invertible sheaf on W such that $\mathcal{A}^{\otimes d} \otimes \mathcal{O}_W(-D_i)$ is generated by global sections for $i = 1, 2$. We put $n = \dim W$. We take smooth divisors

$$
H_1^{(1)}, \ldots, H_n^{(1)}, H_1^{(2)}, \ldots, H_n^{(2)}
$$

on W in general position such that $\mathcal{A}^{\otimes d} = \mathcal{O}_W(D_i + H_j^{(i)})$ for $1 \leq j \leq n$ and $i = 1, 2$. Let $Z_j^{(i)}$ be the cyclic cover associated to $\mathcal{A}^{\otimes d} = \mathcal{O}_W(D_i + H_j^{(i)})$ for $1 \leq j \leq n$ and $i = 1, 2$ (see 2.3.37). Then W' is the normalization of

$$
\left(Z_1^{(1)} \times_W \cdots \times_W Z_n^{(1)} \right) \times_W \left(Z_1^{(2)} \times_W \cdots \times_W Z_n^{(2)} \right).
$$

For the details, see [EsV, 3.19. Lemma] and [Vi7, Lemma 2.5] (see also Lemma 2.3.38). Let $S_j^{(i)}$ be the cyclic cover of V associated to $(f^*\mathcal{A})^{\otimes d} = \mathcal{O}_V(f^*D_i + f^*H_j^{(i)})$. Then we define V' as the normalization of

$$
\left(S_1^{(1)} \times_V \cdots \times_V S_n^{(1)} \right) \times_V \left(S_1^{(2)} \times_V \cdots \times_V S_n^{(2)} \right).
$$

Note that $\rho : V' \to V$ is a finite flat morphism between smooth quasi-projective varieties. Since $V \times_W W' \to V$ is finite and flat and V is smooth, $V \times_W W'$ is Cohen–Macaulay (see, for example, [KolM, Corollary 5.5]). By construction, we can easily see that $V \times_W W'$ is smooth in codimension one. Therefore, $V \times_W W'$ is normal. Since ρ factors through $V \times_W W'$ by construction, we see that $V' = V \times_W W'$ by Zariski's main theorem. $\qquad\square$

Remark 3.3.7 In the proof of Lemma 3.3.6, let S be any simple normal crossing divisor on V. Then we can choose the ramification locus Σ of τ such that $f^*P \not\subset S$ for any irreducible component P of Σ and that $f^*\Sigma + S$ is a simple normal crossing divisor on V. If we choose Σ as above, then we obtain that ρ^*S is a simple normal crossing divisor on V'.

The following lemma is also an application of Lemma 3.3.3.

Lemma 3.3.8 ([Vi3, Lemma 5.4]) *Let $f : V \to W$ be a projective surjective morphism between smooth quasi-projective varieties. Let k be a positive integer and let k' be any multiple of k with $k' \geq 2$. Assume that we have an inclusion*

$$\mathcal{H} \hookrightarrow (f_* \omega_{V/W}^{\otimes k})^{**}$$

for some ample invertible sheaf $\mathcal{H}$ on W. Then there exists a finite surjective morphism $\tau : W' \to W$ from a smooth quasi-projective variety W' such that $V' = V \times_W W'$ is a smooth quasi-projective variety with the following properties:

(i) $\tau^* f_* \omega_{V/W}^{\otimes v} = f'_* \omega_{V'/W'}^{\otimes v}$ for every positive integer v, and
(ii) there exists an ample invertible sheaf $\mathcal{H}'$ on W' such that

$$f'_* \omega_{V'/W'}^{\otimes k'} \otimes \mathcal{H}'^{\otimes -2}$$

is weakly positive.

$$
\begin{array}{ccc}
V & \longleftarrow & V' \\
{\scriptstyle f}\downarrow & & \downarrow{\scriptstyle f'} \\
W & \xleftarrow{\ \tau\ } & W'
\end{array}
$$

Remark 3.3.9 We note that the statement of Lemma 3.3.8 is slightly different from [Vi3, Lemma 5.4].

Proof of Lemma 3.3.8 By the natural map

$$\mathcal{H}^{\otimes a} \to \widehat{S}^a(f_* \omega_{V/W}^{\otimes k}) \to \widehat{S}^1(f_* \omega_{V/W}^{\otimes ak}) = (f_* \omega_{V/W}^{\otimes ak})^{**},$$

we may assume that $k = k' > 1$. By taking blow-ups, if necessary, we may assume that there exist an invertible sheaf $\mathcal{N}$ on V and a simple normal crossing divisor $\sum_j \overline{E}_j$ on V, where $\overline{E}_j$ is smooth for every j and $\overline{E}_i \neq \overline{E}_j$ for $i \neq j$, with

$$\mathcal{N} = \operatorname{Im}\left(f^* f_* \omega_{V/W}^{\otimes k} \to \omega_{V/W}^{\otimes k}\right),$$

$$\mathcal{N} \otimes \mathcal{O}_V(\sum_j \overline{\mu}_j \overline{E}_j) = \omega_{V/W}^{\otimes k},$$

$$\mathcal{N} = f^* \mathcal{H} \otimes \mathcal{O}_V(\sum_j \overline{\nu}_j \overline{E}_j),$$

such that $\overline{\nu}_j \geq 0$ if $\overline{E}_j$ is not f-exceptional. We take a nonempty Zariski open set U' of W such that $f_* \omega_{V/W}^{\otimes k}$ is locally free on U' and f is smooth over U'. By shrinking U', we may assume that $E_j = \overline{E}_j|_{f^{-1}(U')}$ is dominant onto U' if $E_j \neq 0$. We put

$$\mu_j = \begin{cases} \overline{\mu}_j & \text{if } E_j \neq 0 \\ 0 & \text{if } E_j = 0 \end{cases} \quad \text{and} \quad \nu_j = \begin{cases} \overline{\nu}_j & \text{if } E_j \neq 0 \\ 0 & \text{if } E_j = 0. \end{cases}$$

We take a large integer b such that $b > \nu_j$ for all j. We take a general finite flat cover $\tau : W' \to W$ from a smooth quasi-projective variety W' such that $V' = V \times_W W'$ is smooth, $\tau^* f_* \omega_{V/W}^{\otimes m} = f'_* \omega_{V'/W'}^{\otimes m}$ for every $m \geq 1$, and $\tau^* \mathcal{H} \simeq \mathcal{B}^{\otimes 2(bk+1)}$ for some ample invertible sheaf $\mathcal{B}$ on W' by Lemma 3.3.6 (see also Remark 3.3.7). We put $\mathcal{A} = \mathcal{B}^{\otimes 2}$. For simplicity of notation, we may assume that $W = W'$ and that $\mathcal{H} \simeq \mathcal{B}^{\otimes 2(bk+1)} = \mathcal{A}^{\otimes bk+1}$.

From now on, we will prove that $f_* \omega_{V/W}^{\otimes k} \otimes \mathcal{H}'^{\otimes -2}$ is weakly positive, where $\mathcal{H}' = \mathcal{B}^{\otimes k-1}$. By Theorem 3.3.5, $f_* \omega_{V/W}^{\otimes k}$ is weakly positive. Therefore, there exists some $\nu > 0$ such that

$$\widehat{S}^{\nu(b-1)}(f_* \omega_{V/W}^{\otimes k}) \otimes \mathcal{A}^{\otimes \nu}$$

is generically generated by global sections. By Lemma 3.1.12 (i), we may assume that f is flat and that $f_* \omega_{V/W}^{\otimes \eta}$ is locally free for every $\eta \leq \nu(b-1)k$ by replacing W with $W \setminus \Sigma$ for some suitable closed subset Σ of codimension ≥ 2. Of course, we have

$$\widehat{S}^{\nu(b-1)}(f_* \omega_{V/W}^{\otimes k}) = S^{\nu(b-1)}(f_* \omega_{V/W}^{\otimes k}).$$

We put $\mathcal{L} = \omega_{V/W} \otimes f^* \mathcal{A}^{\otimes -1}$, $N = bk$, and

$$\mathcal{M} = \mathcal{L}^{\otimes N} \otimes \mathcal{O}_V(-\sum_j (b\mu_j + \nu_j)\overline{E}_j).$$

Then we have

$$\mathcal{L}^{\otimes N} = \mathcal{M} \otimes \mathcal{O}_V(\sum_j (b\mu_j + \nu_j)\overline{E}_j).$$

By construction, there is an effective divisor F on V such that $\text{Supp} F \subset V \setminus f^{-1}(U')$ and that

$$\mathcal{O}_V(F) = \omega_{V/W}^{\otimes k} \otimes \mathcal{O}_V(-\sum_j (\mu_j + \nu_j)\overline{E}_j) \otimes f^* \mathcal{H}^{\otimes -1}$$

holds. Thus we can check that

$$\mathcal{M} = \left(\omega_{V/W}^{\otimes k} \otimes \mathcal{O}_V(-\sum_j \mu_j \overline{E}_j) \right)^{\otimes b-1} \otimes f^* \mathcal{A} \otimes \mathcal{O}_V(F).$$

The natural maps

$$f^* \widehat{S}^{\nu(b-1)}(f_* \omega_{V/W}^{\otimes k}) \longrightarrow \left(\omega_{V/W}^{\otimes k} \otimes \mathcal{O}_V(-\sum_j \mu_j \overline{E_j}) \right)^{\otimes \nu(b-1)}$$
$$\longrightarrow \mathcal{M}^{\otimes \nu} \otimes f^* \mathcal{A}^{\otimes -\nu}$$

are surjective on $f^{-1}(U')$. Therefore,

$$f^* \left(\widehat{S}^{\nu(b-1)}(f_* \omega_{V/W}^{\otimes k}) \otimes \mathcal{A}^{\otimes \nu} \right) \to \mathcal{M}^{\otimes \nu}$$

is surjective on $f^{-1}(U')$. Thus the assumptions of Lemma 3.3.3 are satisfied, that is, $\mathcal{M}^{\otimes \nu}$ is generated by global sections over $f^{-1}(U)$ for some nonempty Zariski open set U of W. By the choice of b, we have

$$\left\lfloor \frac{(k-1)(b\mu_j + \nu_j)}{bk} \right\rfloor \leq \mu_j + \left\lfloor \frac{\nu_j}{b} \right\rfloor = \mu_j$$

for every j. This means that the sheaf $\omega_{V/W} \otimes \mathcal{L}^{(k-1)}$ contains $\mathcal{N} \otimes f^* \mathcal{A}^{\otimes -(k-1)}$ on $f^{-1}(U')$, where

$$\mathcal{L}^{(k-1)} = \mathcal{L}^{\otimes k-1} \otimes \mathcal{O}_V \left(-\sum_j \left\lfloor \frac{(k-1)(b\mu_j + \nu_j)}{bk} \right\rfloor \overline{E_j} \right).$$

We put $\mathcal{H}' = \mathcal{B}^{\otimes k-1}$ as above. In this case, $\mathcal{H}'^{\otimes 2} = \mathcal{A}^{\otimes k-1}$. Then the inclusion

$$f_*(\omega_{V/W} \otimes \mathcal{L}^{(k-1)}) \to f_* \omega_{V/W}^{\otimes k} \otimes \mathcal{H}'^{\otimes -2}$$

is an isomorphism on U' since $f_* \mathcal{N} = f_* \omega_{V/W}^{\otimes k}$ by construction. Thus, $f_* \omega_{V/W}^{\otimes k} \otimes \mathcal{H}'^{\otimes -2}$ is weakly positive by Lemma 3.1.12 (ii) since $f_*(\omega_{V/W} \otimes \mathcal{L}^{(k-1)})$ is weakly positive by Lemma 3.3.3. $\qquad \square$

As an application of Lemma 3.3.8, we have:

Proposition 3.3.10 ([Vi4, Proposition 3.4]) *Let $f : V \to W$ be a projective surjective morphism between smooth quasi-projective varieties. Let $\mathcal{H}$ be an ample invertible sheaf on W and let $\mathcal{M}$ be any invertible sheaf on W. Let k be a positive integer and let k' be any multiple of k with $k' \geq 2$. Assume that we have an inclusion $\mathcal{H} \hookrightarrow (f_* \omega_{V/W}^{\otimes k})^{**}$. Then*

$$\widehat{S}^\gamma(f_* \omega_{V/W}^{\otimes k'}) \otimes \mathcal{M}^{\otimes -1}$$

is weakly positive for every large positive integer γ. In particular, $f_* \omega_{V/W}^{\otimes k'}$ is big.

Proof By Lemma 3.3.8, there exist a finite cover $\tau : W' \to W$ and an ample invertible sheaf $\mathcal{H}'$ on W' such that $\tau^*(f_* \omega_{V/W}^{\otimes k'}) \otimes \mathcal{H}'^{\otimes -2}$ is weakly positive. Therefore,

$$\widehat{S}^\gamma(\tau^*(f_*\omega_{V/W}^{\otimes k'})\otimes\mathcal{H}'^{\otimes-2})\otimes\mathcal{H}'^{\otimes\gamma} = \widehat{S}^\gamma(\tau^*(f_*\omega_{V/W}^{\otimes k'})\otimes\mathcal{H}'^{\otimes-1})$$

is generically generated by global sections for every large positive integer γ (see Lemma 3.1.3). On the other hand, for every large positive integer γ, $\tau^*\mathcal{M}^{\otimes-1}\otimes\mathcal{H}'^{\otimes\gamma}$ has a nontrivial global section. Thus, $\widehat{S}^\gamma(\tau^*f_*\omega_{V/W}^{\otimes k'}\otimes\mathcal{H}'^{\otimes-1})$ is a subsheaf of $\tau^*(\widehat{S}^\gamma(f_*\omega_{V/W}^{\otimes k'})\otimes\mathcal{M}^{\otimes-1})$. Note that the inclusion

$$\widehat{S}^\gamma(\tau^*f_*\omega_{V/W}^{\otimes k'}\otimes\mathcal{H}'^{\otimes-1}) \hookrightarrow \tau^*(\widehat{S}^\gamma(f_*\omega_{V/W}^{\otimes k'})\otimes\mathcal{M}^{\otimes-1}) \qquad (3.3.1)$$

is an isomorphism at the generic point of W. As we saw above, there exists a generically surjective morphism

$$\bigoplus_{\text{finite}}\mathcal{O}_{W'} \to \widehat{S}^\gamma(\tau^*(f_*\omega_{V/W}^{\otimes k'})\otimes\mathcal{H}'^{\otimes-1}). \qquad (3.3.2)$$

By combining (3.3.1) with (3.3.2), we have a generically surjective morphism

$$\bigoplus_{\text{finite}}\mathcal{O}_{W'} \to \tau^*(\widehat{S}^\gamma(f_*\omega_{V/W}^{\otimes k'})\otimes\mathcal{M}^{\otimes-1}).$$

This implies that

$$\widehat{S}^\gamma(f_*\omega_{V/W}^{\otimes k'})\otimes\mathcal{M}^{\otimes-1}$$

is weakly positive for every large positive integer γ by Lemma 3.1.12 (ii) and (v). $\square$

Remark 3.3.11 Viehweg's original proof of Proposition 3.3.10 uses Theorem 3.1.13. So the above proof is simpler than Viehweg's.

The following theorem is the main theorem of this section.

Theorem 3.3.12 ([Vi4, Theorem 3.5]) *Let* $f : V \to W$ *be a surjective morphism between smooth projective varieties with connected fibers. Assume that* f *is semistable in codimension one. We further assume that*

$$\kappa(W, \widehat{\det}(f_*\omega_{V/W}^{\otimes k})) = \dim W$$

for some positive integer k. *Let* M *be any invertible sheaf on* W *and let* k' *be any multiple of* k *with* $k' \geq 2$. *Then we obtain that*

$$\widehat{S}^\gamma(f_*\omega_{V/W}^{\otimes k'})\otimes\mathcal{M}^{\otimes-1}$$

is weakly positive for every large and divisible positive integer γ. *In particular,* $f_*\omega_{V/W}^{\otimes k'}$ *is big.*

Proof Let $\mathcal{H}$ be an ample invertible sheaf on W. By Kodaira's lemma (see Lemma 2.3.27), we can find $a > 0$ such that $\mathcal{H}$ is contained in $\widehat{\det}(f_*\omega_{V/W}^{\otimes k})^{\otimes a}$. Let U be a

Zariski open set of W such that $\text{codim}_W(W \setminus U) \geq 2$, f is semistable over U, and $f_*\omega_{V/W}^{\otimes k}$ is a locally free sheaf on U. We put $r = \text{rank}(f_*\omega_{V/W}^{\otimes k})|_U$. Then we have an inclusion of $\det(f_*\omega_{V/W}^{\otimes k})|_U$ into $((f_*\omega_{V/W}^{\otimes k})|_U)^{\otimes r}$. Therefore, $\mathcal{H}$ can be seen as a subsheaf of $(f_*\omega_{V/W}^{\otimes k})^{\otimes s}$ for $s = ra$ on U. Let $f^{(s)} : V^{(s)} \to W$ be a desingularization of the s-fold fiber product $V \times_W V \times_W \cdots \times_W V$. Then

$$f_*^{(s)}\omega_{V^{(s)}/W}^{\otimes k} = (f_*\omega_{V/W}^{\otimes k})^{\otimes s}$$

holds on U (see Lemma 3.1.20 and Corollary 3.1.23). Thus, we have $\mathcal{H} \hookrightarrow (f_*^{(s)}\omega_{V^{(s)}/W}^{\otimes k})^{**}$. By Proposition 3.3.10, we obtain that

$$\widehat{S}^\nu(f_*^{(s)}\omega_{V^{(s)}/W}^{\otimes k'}) \otimes \mathcal{M}^{\otimes -1} = \widehat{S}^\nu(((f_*\omega_{V/W}^{\otimes k'})^{\otimes s})^{**}) \otimes \mathcal{M}^{\otimes -1}$$

is weakly positive for every large positive integer ν. Thus $\widehat{S}^{\nu s}(f_*\omega_{V/W}^{\otimes k'}) \otimes \mathcal{M}^{\otimes -1}$ is also weakly positive for every large positive integer ν by Lemma 3.1.12 (ii). $\quad\square$

We close this section with an important remark on weakly semistable morphisms.

Remark 3.3.13 The assumption that $f : V \to W$ is semistable in codimension one in Theorem 3.3.12 can be replaced by the assumption that $f : V \to W$ decomposes as follows:

$$f : V \xrightarrow{\delta} V^\dagger \xrightarrow{f^\dagger} W,$$

where δ is a resolution of singularities and $f^\dagger : V^\dagger \to W$ is weakly semistable. Since $f_*\omega_{V/W}^{\otimes k'} = f_*^\dagger\omega_{V^\dagger/W}^{\otimes k'}$, we may assume that $V = V^\dagger$ for the proof of Theorem 3.3.12. By induction on s, we see that V^s has only Gorenstein singularities by the flat base change theorem [Ve, Theorem 2] (see also [Har2], [Co], and so on).

$$
\begin{array}{ccc}
V^{s-1} & \xleftarrow{\ p\ } & V^s \\
{\scriptstyle f^{s-1}}\downarrow & \ \ {\scriptstyle f^s}\swarrow\ \ \downarrow{\scriptstyle q} \\
W & \xleftarrow{\ f\ } & V
\end{array}
$$

Since $f : V \to W$ is weakly semistable, we can easily see that V^s is normal and is local analytically isomorphic to a toric variety by induction on s. Thus, V^s has only rational Gorenstein singularities and is flat over W. Therefore, $f_*^{(s)}\omega_{V^{(s)}/W}^{\otimes m} = f_*^s\omega_{V^s/W}^{\otimes m}$ is a reflexive sheaf for every positive integer m. By the flat base change theorem [Ve, Theorem 2] (see also [Har2], [Co], and so on), $\omega_{V^s/V} \simeq p^*\omega_{V^{s-1}/W}$. Therefore, we have

$$
\begin{aligned}
f_*^s \omega_{V^s/W}^{\otimes m} &\simeq f_*^{s-1} p_* (p^* \omega_{V^{s-1}/W}^{\otimes m} \otimes q^* \omega_{V/W}^{\otimes m}) \\
&\simeq f_*^{s-1} (\omega_{V^{s-1}/W}^{\otimes m} \otimes p_* q^* \omega_{V/W}^{\otimes m}) \\
&\simeq f_*^{s-1} (\omega_{V^{s-1}/W}^{\otimes m} \otimes (f^{s-1})^* f_* \omega_{V/W}^{\otimes m}) \\
&\simeq \left(f_* \omega_{V/W}^{\otimes m} \otimes (f_*^{s-1} \omega_{V^{s-1}/W}^{\otimes m}) \right)^{**} \\
&\simeq \left(\bigotimes^s f_* \omega_{V/W}^{\otimes m} \right)^{**}
\end{aligned}
$$

by the flat base change theorem (see [Har3, Chap. III, Proposition 9.3]) and the projection formula for every positive integer m and every positive integer s by induction on s. Therefore, the proof of Theorem 3.3.12 also works in this situation.

3.4   Some Further Developments

In this section, we see some further developments and related topics without proof. In Sect. 3.4.1, we will explain some analytic results by Păun–Takayama (see [PăT] and [HPS]). In Sect. 3.4.2, we will look briefly at Nakayama's ω-sheaves and $\widehat{\omega}$-sheaves (see [N4, Chap. V]). In Sect. 3.4.3, we will explain some relatively new results by Dutta–Murayama on twisted weak positivity and effective generation (see [DM]). Of course, our choice of topics is biased and reflects the author's personal taste.

3.4.1   *Singular Hermitian Metrics on Torsion-Free Coherent Sheaves by Păun–Takayama*

Let us recall the following special case of Nakayama's definition of weak positivity for torsion-free coherent sheaves (see [N4, Chap. V, 3.20. Definition]).

Definition 3.4.1 (*see* [PăT, *Definition 2.5.1*]) Let $\mathcal{F}$ be a torsion-free coherent sheaf on a smooth projective variety Y. We say that $\mathcal{F}$ is *weakly positive at a point* $y \in Y$ if for every positive integer α and every ample invertible sheaf $\mathcal{H}$ on Y there exists a positive integer β such that $\widehat{S}^{\alpha\beta}(\mathcal{F}) \otimes \mathcal{H}^{\otimes\beta}$ is generated by global sections at y.

In [PăT], Păun and Takayama introduced the notion of singular hermitian metrics on torsion-free coherent sheaves (see [PăT, Definition 2.4.1]) and obtained the following statement.

Theorem 3.4.2 ([PăT, Theorem 2.5.2]) *Let $\mathcal{E}$ be a torsion-free coherent sheaf on a smooth projective variety Y. Suppose that $\mathcal{E}$ admits a singular hermitian metric h with semipositive curvature in the sense of Păun–Takayama (see [PăT, Definition*

2.4.1]). Then $\mathcal{E}$ is weakly positive at y for every $y \in Y$ where $\mathcal{E}$ is locally free around y and $\det h(y) < \infty$.

For the precise definitions and basic properties of singular hermitian metrics on torsion-free coherent sheaves, we recommend the reader to see [PăT, Sect. 2].

By using the Ohsawa–Takegoshi L^2-extension theorem, Păun and Takayama also established the following important result.

Theorem 3.4.3 ([PăT, Theorem 1.1]) *Let $f : X \to Y$ be a surjective morphism between smooth projective varieties with connected fibers. Let m be a positive integer such that $f_*\omega_{X/Y}^{\otimes m} \neq 0$. Then the torsion-free coherent sheaf $f_*\omega_{X/Y}^{\otimes m}$ admits a singular hermitian metric with semipositive curvature in the sense of Păun–Takayama (see [PăT, Definition 2.4.1]).*

We can directly recover Viehweg's weak positivity theorem (see Theorems 3.2.5 and 3.3.5) by combining Theorems 3.4.3 with 3.4.2 when general fibers are connected. For the details of Theorem 3.4.3 and some important generalizations, we strongly recommend the reader to see [PăT], [HPS]. We note that [HPS] is very accessible to algebraic geometers.

3.4.2   Nakayama's Theory of ω-Sheaves, and so on

Let us quickly recall the notion of ω-sheaves and $\widehat{\omega}$-sheaves introduced by Nakayama since it plays an important role in [F13]. We only treat projective varieties for simplicity, although Nakayama introduced the notion of ω-sheaves for complex analytic spaces.

Definition 3.4.4 ([N4, Chap. V, 3.8. Definition]) A coherent sheaf $\mathcal{F}$ on a projective variety Y is called an ω-*sheaf* if there exists a projective morphism $f : X \to Y$ from a smooth projective variety X such that $\mathcal{F}$ is a direct summand of $R^i f_*\omega_X$ for some i.

Definition 3.4.5 ([N4, Chap. V, 3.16. Definition]) A coherent sheaf $\mathcal{F}$ on a normal projective variety Y is called an $\widehat{\omega}$-*sheaf* if there exist an ω-sheaf $\mathcal{G}$ and a generically isomorphic injection $\mathcal{G} \hookrightarrow \mathcal{F}^{**}$ into the double dual $\mathcal{F}^{**}$ of $\mathcal{F}$.

In [N4], Nakayama also introduced the notion of ω-*bigness* (see [N4, Chap. V, 3.16. Definition]). Then he discussed some sophisticated generalizations of Viehweg's weak positivity theorem. By combining them with his theory of numerical dimensions (see [N4, Chap. V], [Leh], [Ec], [Les], and [F18]), he established various applications to the Iitaka conjecture and the abundance conjecture. For the details, see [N4, Chap. V].

Remark 3.4.6 In [F19], we introduce the notion of *pure-ω-sheaves* and *mixed-ω-sheaves* following Nakayama's theory of ω-sheaves. Our approach in [F19] is more natural than [N4, Chap. V] from the mixed Hodge theoretic viewpoint. Unfortunately, however, [F19] and [N4, Chap. V] are not completely compatible.

3.4.3   Twisted Weak Positivity and Effective Generation By Dutta–Murayama

Here, we will freely use the standard notation of the minimal model program as in [F7] and [F12]. In [DM], Dutta and Murayama established the following generalization of Viehweg's weak positivity theorem.

Theorem 3.4.7 (Twisted weak positivity, [DM, Theorem D]) *Let $f : Y \to X$ be a surjective morphism between normal projective varieties with connected fibers such that X is Gorenstein. Let Δ be an effective $\mathbb{R}$-Cartier $\mathbb{R}$-divisor on Y such that (Y, Δ) is lc and $k(K_X + \Delta)$ is $\mathbb{R}$-linearly equivalent to a Cartier divisor for some positive integer k. Then the sheaf*

$$f_* \mathcal{O}_Y(k(K_{Y/X} + \Delta))$$

is weakly positive in the sense of Definition 3.4.8 below.

Definition 3.4.8 Let $\mathcal{F}$ be a torsion-free coherent sheaf on a normal projective variety X. We say that $\mathcal{F}$ is *weakly positive* if there exists a nonempty Zariski open set U such that for every positive integer α and every ample invertible sheaf $\mathcal{H}$, there is a positive integer β such that

$$\widehat{S}^{\alpha\beta}(\mathcal{F}) \otimes \mathcal{H}^{\otimes\beta}$$

is generated by global sections on U.

We note that the Zariski open set U is independent of α and $\mathcal{H}$ in Definition 3.4.8. Therefore, the condition in Definition 3.4.8 is much stronger than that in Definition 3.1.1. In this book, we adopted Definition 3.1.1 for weak positivity since it is sufficient for our applications and is easier to check than Definition 3.4.8.

In [DM], the authors also obtained a generalization of Remark 3.2.6.

Theorem 3.4.9 (Effective generation, [DM, Theorem E]) *Let $f : Y \to X$ be a surjective morphism from a smooth projective variety Y to a normal Gorenstein projective variety X with connected fibers. Let Δ be an $\mathbb{R}$-divisor on Y such that $\operatorname{Supp}\Delta$ is a simple normal crossing divisor and that the coefficients of Δ_{hor} are in $(0, 1]$. Consider a Cartier divisor P on Y such that $P \sim_{\mathbb{R}} k(K_Y + \Delta)$ for some positive integer k. Let U be the intersection of $U(f, \Delta)$ with the largest Zariski open set over which $f_* \mathcal{O}_Y(P)$ is locally free. Let $\mathcal{A}$ be an ample invertible sheaf on X such that $|\mathcal{A}|$ is free. Then the sheaf*

$$\left(\bigotimes^{s} f_* \mathcal{O}_Y(k(K_{Y/X} + \Delta)) \right)^{**} \otimes \left(\omega_X \otimes \mathcal{A}^{\otimes(\dim X+1)} \right)^{\otimes l}$$

is generated by global sections on U for all integers $l \geq k$ and $s \geq 1$.

We note that $U(f, \Delta)$ in Theorem 3.4.9 is the largest Zariski open set of X such that $U(f, \Delta)$ is contained in the smooth locus of X, $f : f^{-1}(U(f, \Delta)) \to U(f, \Delta)$ is smooth, and the fibers $f^{-1}(x)$ intersect each component of Δ transversally for all closed points $x \in U(f, \Delta)$.

The above results are generalizations and continuations of [PopS] and Sect. 3.2. We strongly recommend the interested reader to see [DM]. For the original treatment of the twisted weak positivity theorem mainly due to Viehweg and Campana, see [Cam, 4.4. Twisted weak positivity] and [F15] (see also [F9, Sect. 8]). For some geometric applications and related topics, we recommend the reader to see [F9], [F15], [F19], [Has2], [KovP], and so on.

Chapter 4
Iitaka–Viehweg Conjecture for Some Special Cases

Let us treat the Iitaka conjecture and the Viehweg conjecture in some special but interesting cases. Our treatment is relatively classical and is mainly due to Viehweg, although it depends on [AbK] and [BCHM].

Let $f : X \to Y$ be a surjective morphism between smooth projective varieties with connected fibers. In this chapter, we treat the Iitaka conjecture in the following three cases:

- Y is of general type,
- the geometric generic fiber $X_{\overline{\eta}}$ is of general type, and
- general fibers of $f : X \to Y$ are elliptic curves.

In Sect. 4.2, we prove the Iitaka conjecture under the assumption that Y is of general type. This case easily follows from Fujita's lemma (see Lemma 4.1.4) and Viehweg's weak positivity of $f_*\omega_{X/Y}^{\otimes m}$. It is one of the easiest cases of the Iitaka conjecture. In Sect. 4.1, we explain that the (generalized) Iitaka conjecture follows from the Viehweg conjecture in details. It is mainly due to Viehweg. Our treatment is slightly different from the traditional one. This is because we use the weak semistable reduction theorem by Abramovich–Karu (see [AbK]). In Sect. 4.3, we prove the Viehweg conjecture under the assumption that the geometric generic fiber is of general type. We need no deep results coming from the theory of variations of Hodge structure. Therefore, we believe that our proof is more accessible than [Kaw6] and [Kol3]. We use the weak semistable reduction theorem and the existence of relative canonical models for fiber spaces whose geometric generic fiber is of general type by Birkar–Cascini–Hacon–McKernan (see [BCHM]). We note that some of our results in this section are sharper than the known results (see [Kaw6], [Kol3], and [Vi6]). A key result in this section is that $f_*\omega_{X/Y}^{\otimes m}$ is locally free for every positive integer m if f is weakly semistable and the geometric generic fiber of f is of general type, which was first obtained in [F10] and [F11]. In Sect. 4.4, we explain the Viehweg conjecture for elliptic fibrations. This section is not self-contained. We use some well-known

© The Author(s), under exclusive license to Springer Nature Singapore Pte Ltd. 2020
O. Fujino, *Iitaka Conjecture*, SpringerBriefs in Mathematics,
https://doi.org/10.1007/978-981-15-3347-1_4

results on elliptic fibrations and moduli spaces of elliptic curves without proof. We need a result in Sect. 4.4 for the proof of $\overline{C}_{n,n-1}$ in Chap. 5.

4.1   From Viehweg's Conjecture to Iitaka's Conjecture

This section is a slight reformulation of [Vi3, Sect. 7]. We prove that Viehweg's conjecture (see Conjecture 1.2.6) implies the generalized Iitaka conjecture (see Conjecture 1.2.5).

First, let us recall the definition of Viehweg's variation.

Definition 4.1.1 (*Viehweg's variation*) Let $f : X \to Y$ be a surjective morphism between normal projective varieties. Let $K(\supset \mathbb{C})$ be an algebraically closed field contained in $\overline{\mathbb{C}(Y)}$ such that there is a smooth projective variety V defined over K and that $V \times_{\operatorname{Spec} K} \operatorname{Spec} \overline{\mathbb{C}(Y)}$ and $X \times_Y \operatorname{Spec} \overline{\mathbb{C}(Y)}$ are birational. The minimum of trans.$\deg_{\mathbb{C}} K$ for all such K is called the *variation* of f and is denoted by $\operatorname{Var}(f)$. We have $0 \le \operatorname{Var}(f) \le \dim Y$.

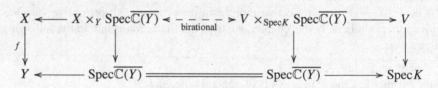

Next, we recall Viehweg's Conjecture $Q_{n,m}$ (see Conjecture 1.2.6).

Conjecture 4.1.2 (*Viehweg's Conjecture $Q_{n,m}$*) *Let $f : X \to Y$ be a surjective morphism between smooth projective varieties with connected fibers such that $\kappa(X_{\overline{\eta}}) \ge 0$, where $X_{\overline{\eta}}$ is the geometric generic fiber of $f : X \to Y$, and that $\dim X = n$ and $\dim Y = m$. Assume that $\operatorname{Var}(f) = \dim Y$. Then $f_* \omega_{X/Y}^{\otimes k}$ is big for some positive integer k.*

Remark 4.1.3 If $\kappa(X_{\overline{\eta}}) = -\infty$ in Conjecture 4.1.2, then $f_* \omega_{X/Y}^{\otimes k} = 0$ for every positive integer k.

We prepare Fujita's easy but important lemma (see [Ft1, Proposition 1]).

Lemma 4.1.4 (*Fujita's lemma*) *Let $f : X \to Y$ be a projective surjective morphism between normal projective varieties with connected fibers. Let $\mathcal{L}$ be an invertible sheaf on X and let M be an invertible sheaf on Y such that $\kappa(Y, M) = \dim Y$ and $\kappa(X, \mathcal{L}^{\otimes a} \otimes f^* M^{\otimes -b}) \ge 0$ for some positive integers a and b. Then we have*

$$\kappa(X, \mathcal{L}) = \kappa(X_{\overline{\eta}}, \mathcal{L}|_{X_{\overline{\eta}}}) + \kappa(Y, M),$$

where $X_{\overline{\eta}}$ is the geometric generic fiber of $f : X \to Y$.

Proof By Iitaka's easy addition formula (see Lemma 2.3.31), we have

$$\kappa(X, \mathcal{L}) \leq \dim Y + \kappa(X_{\overline{\eta}}, \mathcal{L}|_{X_{\overline{\eta}}}).$$

Therefore, it is sufficient to prove

$$\kappa(X, \mathcal{L}) \geq \kappa(Y, \mathcal{M}) + \kappa(X_{\overline{\eta}}, \mathcal{L}|_{X_{\overline{\eta}}}).$$

By Kodaira's lemma (see Lemma 2.3.27), we may assume that $\mathcal{M}$ is ample. We may further assume that $\mathcal{M}$ is very ample, the rational map $\Phi_{|\mathcal{L}|} : X \dashrightarrow V \subset \mathbb{P}^{\dim |\mathcal{L}|}$ gives an Iitaka fibration, and $H^0(X, \mathcal{L} \otimes f^*\mathcal{M}^{\otimes -1}) \neq 0$ by replacing $\mathcal{L}$ and $\mathcal{M}$ with multiplies. An element $\alpha \neq 0$ of $H^0(X, \mathcal{L} \otimes f^*\mathcal{M}^{\otimes -1})$ defines an injection $H^0(Y, \mathcal{M}) \hookrightarrow H^0(X, \mathcal{L})$. Therefore, it gives a projection

$$\mathbb{P}^{\dim |\mathcal{L}|} \dashrightarrow \mathbb{P}^{\dim |\mathcal{M}|}.$$

Hence we obtain the following commutative diagram:

$$
\begin{array}{ccccc}
X & \overset{\Phi_{|\mathcal{L}|}}{\dashrightarrow} & V & \hookrightarrow & \mathbb{P}^{\dim |\mathcal{L}|} \\
{\scriptstyle f}\downarrow & & {\scriptstyle \pi}\downarrow & & \downarrow \\
Y & \underset{\Phi_{|\mathcal{M}|}}{\overset{\sim}{\to}} & W & \hookrightarrow & \mathbb{P}^{\dim |\mathcal{M}|}
\end{array}
$$

By taking suitable resolutions of X and V in the above diagram, we may assume that we have

$$
\begin{array}{ccc}
X & \overset{\rho}{\to} & \widetilde{V} \\
{\scriptstyle f}\downarrow & & \downarrow{\scriptstyle \widetilde{\pi}} \\
Y & = & Y
\end{array}
$$

where $\widetilde{V}$ is a smooth projective variety which is birationally equivalent to V. We take a sufficiently general point y of Y and consider the mapping

$$\rho_y : X_y = f^{-1}(y) \to \widetilde{V}_y = \widetilde{\pi}^{-1}(y).$$

A sufficiently general fiber F of ρ_y is also a sufficiently general fiber of ρ. Therefore, we have $\kappa(F, \mathcal{L}|_F) = 0$. Note that ρ is an Iitaka fibration with respect to $\mathcal{L}$. Thus, we have

$$
\begin{aligned}
\kappa(X_y, \mathcal{L}|_{X_y}) &\leq \kappa(F, \mathcal{L}|_F) + \dim \widetilde{V}_y \\
&= \dim \widetilde{V} - \dim Y \\
&= \kappa(X, \mathcal{L}) - \kappa(Y, \mathcal{M})
\end{aligned}
$$

by Iitaka's easy addition formula (see Lemma 2.3.31). On the other hand, we have $\kappa(X_y, \mathcal{L}|_{X_y}) = \kappa(X_{\overline{\eta}}, \mathcal{L}|_{X_{\overline{\eta}}})$. Therefore, the desired inequality

$$\kappa(X, \mathcal{L}) \geq \kappa(Y, \mathcal{M}) + \kappa(X_{\overline{\eta}}, \mathcal{L}|_{X_{\overline{\eta}}})$$

holds.

The big commutative diagram constructed in Lemma 4.1.5 below will play an important role.

Lemma 4.1.5 *Let $f : X \to Y$ be a surjective morphism between smooth projective varieties with connected fibers. Then we have a commutative diagram*

$$
\begin{array}{ccccccc}
X & \xleftarrow{\alpha} & V & \xleftarrow{\rho} & V' & \xrightarrow{\rho''} & V'' \\
{\scriptstyle f}\downarrow & & {\scriptstyle g}\downarrow & & {\scriptstyle g'}\downarrow & & {\scriptstyle g''}\downarrow \\
Y & \xleftarrow{\beta} & W & \xleftarrow{\tau} & W' & \xrightarrow{\tau''} & W''
\end{array}
$$

satisfying the following properties:

  (i)   *V and W are smooth projective varieties.*
 (ii)   *α and β are birational.*
(iii)   *All g-exceptional divisors are α-exceptional.*
 (iv)   *W'' is a smooth projective variety.*
  (v)   *V'' and W' are normal projective varieties.*
 (vi)   *$\dim W'' = \mathrm{Var}(g'') = \mathrm{Var}(f)$.*
(vii)   *$\tau : W' \to W$ is a generically finite surjective morphism.*
(viii) *V' is a resolution of $W' \times_{W''} V''$ and is a resolution of the main component of $V \times_W W'$ at the same time.*
 (ix)   *$g'' : V'' \to W''$ and $\tau'' : W' \to W''$ have connected fibers and are weakly semistable.*

Proof We divide the proof into several steps.

Step 1 By the flattening theorem (see, for example, [AbO, 3.3. The flattening lemma]), we can find a projective birational morphism $\beta : W \to Y$ from a smooth projective variety W such that $(W \times_Y X)_{\mathrm{main}} \to W$ induced by $\beta : W \to Y$ is flat, where $(W \times_Y X)_{\mathrm{main}}$ is the main component of $W \times_Y X$. Let $V \to (W \times_Y X)_{\mathrm{main}}$ be a projective birational morphism from a smooth projective variety V. Then we have the following commutative diagram:

$$
\begin{array}{ccc}
X & \xleftarrow{\alpha} & V \\
{\scriptstyle f}\downarrow & & {\scriptstyle g}\downarrow \\
Y & \xleftarrow{\beta} & W
\end{array}
$$

satisfying (i), (ii), and (iii).

Step 2 Note that $\mathrm{Var}(f) = \mathrm{Var}(g)$ by definition. Therefore, we can construct the following commutative diagram:

$$
\begin{array}{ccccc}
V & \xleftarrow{\ \rho\ } & V' & \xrightarrow{\ \rho''\ } & V'' \\
{\scriptstyle g}\downarrow & & {\scriptstyle g'}\downarrow & & {\scriptstyle g''}\downarrow \\
W & \xleftarrow[\ \tau\ ]{} & W' & \xrightarrow[\ \tau''\ ]{} & W''
\end{array}
$$

such that V', W', V'', and W'' are smooth projective varieties, g'' is a surjective morphism between smooth projective varieties with connected fibers, $\dim W'' = \mathrm{Var}(g'') = \mathrm{Var}(g) = \mathrm{Var}(f)$, $\tau : W' \to W$ is a generically finite surjective morphism, V' is a resolution of the main component of $V \times_W W'$ and is a resolution of the main component of $V'' \times_{W''} W'$ at the same time. Without loss of generality, we may assume that τ'' has connected fibers.

Step 3 By the weak semistable reduction theorem (see Theorem 2.3.46 and [AbK, Theorem 0.3]), we may assume that $g'' : V'' \to W''$ is weakly semistable by taking the base change by a generically finite surjective morphism $W^\dagger \to W''$ from a smooth projective variety $W^\dagger$. By applying the weak semistable reduction theorem to $\tau'' : W' \to W''$, we may further assume that $\tau'' : W' \to W''$ is also weakly semistable by the base change by a generically finite morphism $W^{\dagger\dagger} \to W''$ from a smooth projective variety $W^{\dagger\dagger}$ (see Lemma 2.3.49 and [AbK, Lemma 6.2]). Then we have a commutative diagram of V, V', V'', W, W' and W'' satisfying properties (iv)–(ix).

Therefore, we have a desired big commutative diagram satisfying properties (i)–(ix). $\qquad\square$

Lemma 4.1.6 *Let $\mathcal{L}$ be an invertible sheaf on Y. Then we have*

$$
\kappa(X, \omega_{X/Y} \otimes f^*\mathcal{L}) \geq \kappa(V, \omega_{V/W} \otimes \mathcal{O}_V(B) \otimes \alpha^* f^*\mathcal{L})
$$

for any effective g-exceptional divisor B on V.

Proof We can write $K_V = \alpha^* K_X + E$ and $K_W = \beta^* K_Y + F$ such that E is an effective α-exceptional divisor and F is an effective β-exceptional divisor. Therefore,

$$
K_{V/W} + B = K_V - g^* K_W + B = \alpha^* K_{X/Y} + E + B - g^* F \leq \alpha^* K_{X/Y} + E + B.
$$

Note that $E + B$ is an effective α-exceptional divisor. Therefore, we obtain

$$
\kappa(X, \omega_{X/Y} \otimes f^*\mathcal{L}) \geq \kappa(V, \omega_{V/W} \otimes \mathcal{O}_V(B) \otimes \alpha^* f^*\mathcal{L})
$$

for any effective g-exceptional divisor B. $\qquad\square$

Lemma 4.1.7 essentially says that Viehweg's conjecture (see Conjecture 1.2.6) implies the generalized Iitaka conjecture (see Conjecture 1.2.5).

Lemma 4.1.7 *Assume that* $\widehat{\det} g''_* \omega_{V''/W''}^{\otimes m}$ *is a big invertible sheaf for some positive integer m. Then we obtain*

$$\kappa(X, \omega_{X/Y} \otimes f^*\mathcal{L}) \geq \kappa(X_{\overline{\eta}}) + \max\{\mathrm{Var}(f), \kappa(Y, \mathcal{L})\}$$

for every invertible sheaf $\mathcal{L}$ *on* Y *with* $\kappa(Y, \mathcal{L}) \geq 0$, *where* $X_{\overline{\eta}}$ *is the geometric generic fiber of* $f : X \to Y$.

Proof We need several steps for the proof of Lemma 4.1.7.

Step 1 By the proof of Theorem 3.3.12 (see also Remark 3.3.13), we have that $g''_* \omega_{V''/W''}^{\otimes k}$ is big for some positive integer k. Therefore, there is a positive integer ν such that there exists a generically isomorphic inclusion

$$\bigoplus \mathcal{H} \hookrightarrow \widehat{S}^\nu(g''_* \omega_{V''/W''}^{\otimes k}),$$

where $\mathcal{H}$ is an ample invertible sheaf on W'', by Lemma 3.1.15. By the nonzero map

$$\widehat{S}^\nu(g''_* \omega_{V''/W''}^{\otimes k}) \to g''_* \omega_{V''/W''}^{\otimes \nu k},$$

we may assume that $g''_* \omega_{V''/W''}^{\otimes k}$ contains an ample Cartier divisor H on W'', that is, $\mathcal{O}_{W''}(H) \subset g''_* \omega_{V''/W''}^{\otimes k}$, by replacing νk with k.

Step 2 We consider the following commutative diagram:

$$
\begin{array}{ccc}
\widetilde{V} & \xrightarrow{\tilde{\rho}} & V'' \\
\downarrow{\scriptstyle \tilde{g}} & & \downarrow{\scriptstyle g''} \\
W' & \xrightarrow{\tau''} & W''
\end{array}
$$

where $\widetilde{V} = W' \times_{W''} V''$. Then we obtain

$$(\tau'')^* g''_* \omega_{V''/W''}^{\otimes k} \simeq \tilde{g}_* \omega_{\widetilde{V}/W'}^{\otimes k}$$

by the flat base change theorem [Ve, Theorem 2] (see also [Har2], [Co], and so on). Note that $\widetilde{V}$ has only rational Gorenstein singularities (see Lemma 2.3.48). This implies that $g'_* \omega_{V'/W'}^{\otimes k} \simeq \tilde{g}_* \omega_{\widetilde{V}/W'}^{\otimes k}$. We note that $\tilde{g}_* \omega_{\widetilde{V}/W'}^{\otimes k}$ is a reflexive sheaf on W'. We obtain that $\tilde{g}_* \omega_{\widetilde{V}/W'}^{\otimes k} (\simeq g'_* \omega_{V'/W'}^{\otimes k})$ contains $(\tau'')^* H$. So we have that $\omega_{\widetilde{V}/W'}^{\otimes k}$ contains $(\tau'' \circ \tilde{g})^* H$.

Step 3 In this step, we will check

$$\kappa(V', \omega_{V'/W'} \otimes \rho^* \alpha^* f^* \mathcal{L}) \geq \kappa(V_{\overline{\eta}}) + \max\{\text{Var}(g), \kappa(W, \beta^* \mathcal{L})\}$$
$$= \kappa(X_{\overline{\eta}}) + \max\{\text{Var}(f), \kappa(Y, \mathcal{L})\},$$

where $V_{\overline{\eta}}$ is the geometric generic fiber of g and $X_{\overline{\eta}}$ is the geometric generic fiber of f.

Since $\omega_{\widetilde{V}/W'}^{\otimes k}$ contains $(\tau'' \circ \widetilde{g})^* H$ and $\kappa(Y, \mathcal{L}) \geq 0$, we obtain

$$\kappa(\widetilde{V}, (\omega_{\widetilde{V}/W'} \otimes \widetilde{g}^* \tau^* \beta^* \mathcal{L})^{\otimes a} \otimes (\tau'' \circ \widetilde{g})^* \mathcal{O}_{W''}(-bH)) \geq 0$$

for some positive integers a and b. Then, by Lemma 4.1.4, we obtain

$$\kappa(V', \omega_{V'/W'} \otimes \rho^* \alpha^* f^* \mathcal{L}) = \kappa(V', \omega_{V'/W'} \otimes g'^* \tau^* \beta^* \mathcal{L})$$
$$= \kappa(\widetilde{V}, \omega_{\widetilde{V}/W'} \otimes \widetilde{g}^* \tau^* \beta^* \mathcal{L})$$
$$= \dim W'' + \kappa(\widetilde{V}_{\overline{w''}}, (\omega_{\widetilde{V}/W'} \otimes \widetilde{g}^* \tau^* \beta^* \mathcal{L})|_{\widetilde{V}_{\overline{w''}}})$$
$$= \dim W'' + \kappa(V''_{\overline{w''}}, \omega_{V''_{\overline{w''}}}) + \kappa(W'_{\overline{w''}}, \tau^* \beta^* \mathcal{L}|_{W'_{\overline{w''}}})$$
$$= \dim W'' + \kappa(V_{\overline{\eta}}) + \kappa(W'_{\overline{w''}}, \tau^* \beta^* \mathcal{L}|_{W'_{\overline{w''}}}).$$

Note that $\widetilde{V}_{\overline{w''}} = W'_{\overline{w''}} \times V''_{\overline{w''}}$, where $\overline{w''}$ is the geometric generic point of W''. Since $\dim W'' = \text{Var}(g)$ and

$$\kappa(W, \beta^* \mathcal{L}) = \kappa(W', \tau^* \beta^* \mathcal{L}) \leq \dim W'' + \kappa(W'_{\overline{w''}}, \tau^* \beta^* \mathcal{L}|_{W'_{\overline{w''}}})$$

by Lemma 2.3.31, we obtain

$$\kappa(V', \omega_{V'/W'} \otimes \rho^* \alpha^* f^* \mathcal{L}) \geq \kappa(V_{\overline{\eta}}) + \max\{\text{Var}(g), \kappa(W, \beta^* \mathcal{L})\}$$
$$= \kappa(X_{\overline{\eta}}) + \max\{\text{Var}(f), \kappa(Y, \mathcal{L})\}.$$

Step 4 Let U be a Zariski open set of W such that g is flat over U and that $\text{codim}_W(W \setminus U) \geq 2$. By restricting

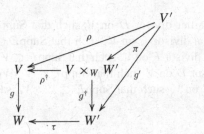

to U, we obtain

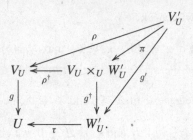

Without loss of generality, we may assume that W'_U is smooth and $\tau : W'_U \to U$ is flat by shrinking U. We note that $\tau : W' \to W$ is generically finite. By the base change theorem (see Lemma 3.1.20), we obtain

$$g'_* \omega^{\otimes l}_{V'_U / W'_U} \hookrightarrow \tau^*(g_* \omega^{\otimes l}_{V_U / U}) \simeq g^\dagger_*((\rho^\dagger)^* \omega^{\otimes l}_{V_U / U}) \tag{4.1.1}$$

for every positive integer l. We note that there exists the following natural inclusion:

$$(\rho^\dagger)^* \omega^{\otimes l}_{V_U / U} \hookrightarrow \pi_* \pi^* (\rho^\dagger)^* \omega^{\otimes l}_{V_U / U} \tag{4.1.2}$$

for every l. Therefore, by (4.1.1) and (4.1.2), we have

$$g'_*(\omega_{V'_U / W'_U} \otimes g'^* \tau^* \beta^* \mathcal{L})^{\otimes l} \hookrightarrow g'_*(\rho^*(\omega_{V_U / U} \otimes g^* \beta^* \mathcal{L})^{\otimes l}). \tag{4.1.3}$$

We note that $\kappa(V', \omega_{V'/W'} \otimes g'^* \tau^* \beta^* \mathcal{L}) = \kappa(V', \omega_{V'/W'} \otimes \rho^* \alpha^* f^* \mathcal{L}) \geq 0$ by Step 3. We fix a sufficiently large and divisible positive integer l such that the linear system $|(\omega_{V'/W'} \otimes g'^* \tau^* \beta^* \mathcal{L})^{\otimes l}|$ gives an Iitaka fibration when $\kappa(V', \omega_{V'/W'} \otimes g'^* \tau^* \beta^* \mathcal{L}) > 0$. If $\kappa(V', \omega_{V'/W'} \otimes g'^* \tau^* \beta^* \mathcal{L}) = 0$, then we take a positive integer such that

$$H^0(V', (\omega_{V'/W'} \otimes g'^* \tau^* \beta^* \mathcal{L})^{\otimes l}) \neq 0.$$

By (4.1.3), we have the following inclusion:

$$H^0(V', (\omega_{V'/W'} \otimes g'^* \tau^* \beta^* \mathcal{L})^{\otimes l})$$
$$\hookrightarrow H^0(V', \rho^*((\omega_{V/W} \otimes g^* \beta^* \mathcal{L})^{\otimes l} \otimes \mathcal{O}_V(D)) \otimes \mathcal{O}_{V'}(E))$$

for some effective Cartier divisor D on V such that $\operatorname{Supp} D \subset V \setminus V_U$ and some effective ρ-exceptional divisor E on V' such that $\operatorname{Supp} E \subset V' \setminus V'_U$. We note that for any effective Weil divisor F on V' such that $\operatorname{Supp} F \subset V' \setminus V'_U$ we can find some effective Cartier divisor D on V such that $\operatorname{Supp} D \subset V \setminus V_U$ and some effective ρ-exceptional divisor E on V' such that $\operatorname{Supp} E \subset V' \setminus V'_U$ satisfying $F \leq \rho^* D + E$. Therefore, we obtain

$$\kappa(V', \omega_{V'/W'} \otimes \rho^* \alpha^* f^* \mathcal{L})$$
$$= \kappa(V', \omega_{V'/W'} \otimes g'^* \tau^* \beta^* \mathcal{L})$$
$$\leq \kappa(V', \rho^*((\omega_{V/W} \otimes g^* \beta^* \mathcal{L})^{\otimes l} \otimes \mathcal{O}_V(D)) \otimes \mathcal{O}_{V'}(E))$$
$$\leq \kappa(V, \omega_{V/W} \otimes \mathcal{O}_V(B) \otimes \alpha^* f^* \mathcal{L})$$

for some effective Cartier divisor B on V such that $\mathrm{Supp}\, B \subset V \setminus V_U$. Note that B is g-exceptional. Therefore, B is α-exceptional.

Thus, by Lemma 4.1.6 and the inequalities obtained above, we obtain

$$\kappa(X, \omega_{X/Y} \otimes f^* \mathcal{L}) \geq \kappa(V, \omega_{V/W} \otimes \mathcal{O}_V(B) \otimes \alpha^* f^* \mathcal{L})$$
$$\geq \kappa(V', \omega_{V'/W'} \otimes \rho^* \alpha^* f^* \mathcal{L})$$
$$\geq \kappa(X_{\overline{\eta}}) + \max\{\mathrm{Var}(f), \kappa(Y, \mathcal{L})\}.$$

This is the desired inequality. □

We will apply Lemma 4.1.7 to algebraic fiber spaces whose geometric generic fiber is of general type and elliptic fibrations in Sects. 4.3 and 4.4, respectively.

Lemma 4.1.8 *Assume that* $\widehat{\det} g''_* \omega^{\otimes m}_{V''/W''}$ *is a big invertible sheaf for some positive integer m. Then we have*

$$\kappa(Y, \widehat{\det}(f_* \omega^{\otimes m}_{X/Y})) \geq \dim W'' = \mathrm{Var}(f).$$

Proof Note that

$$\tau''^* g''_* \omega^{\otimes m}_{V''/W''} = \widetilde{g}_* \omega^{\otimes m}_{\widetilde{V}/W'} = g'_* \omega^{\otimes m}_{V'/W'}$$

(see Step 2 in the proof of Lemma 4.1.7). Therefore, we have

$$\kappa(W', \widehat{\det} g'_* \omega^{\otimes m}_{V'/W'}) = \kappa(W', \widehat{\det} \widetilde{g}_* \omega^{\otimes m}_{V'/W'})$$
$$= \kappa(W', \tau''^* \widehat{\det} g''_* \omega^{\otimes m}_{V''/W''}) = \dim W''.$$

We note that

$$\tau''^* \widehat{\det} g''_* \omega^{\otimes m}_{V''/W''} = \widehat{\det} \tau''^* g''_* \omega^{\otimes m}_{V''/W''}$$

since τ'' is flat. Let $U^{\dagger}$ be a Zariski open set of Y such that $\tau \circ \beta : W' \to Y$ is flat over $U^{\dagger}$ and that $\mathrm{codim}_Y(Y \setminus U^{\dagger}) \geq 2$. We may further assume that $f_* \omega^{\otimes m}_{X_{U^{\dagger}}/U^{\dagger}}$ is locally free. By restricting

$$\begin{array}{ccc}
X & \xleftarrow{\alpha \circ \rho} & V' \\
\downarrow f & & \downarrow g' \\
Y & \xleftarrow{\beta \circ \tau} & W'
\end{array}$$

to $U^\dagger$, we obtain

$$
\begin{array}{ccc}
X_{U^\dagger} & \xleftarrow{\ \alpha \circ \rho\ } & V'_{U^\dagger} \\
{\scriptstyle f}\downarrow & & \downarrow{\scriptstyle g'} \\
U^\dagger & \xleftarrow{\ \beta \circ \tau\ } & W'_{U^\dagger}.
\end{array}
$$

Without loss of generality, we may further assume that $W'_{U^\dagger}$ is smooth since $\beta \circ \tau :$ $W' \to Y$ is generically finite. By the base change theorem (see Lemma 3.1.20), we obtain a generically isomorphic inclusion

$$
g'_* \omega^{\otimes m}_{V'_{U^\dagger}/W'_{U^\dagger}} \hookrightarrow (\beta \circ \tau)^* (f_* \omega^{\otimes m}_{X_{U^\dagger}/U^\dagger}).
$$

This implies that there exists an inclusion of invertible sheaves:

$$
\widehat{\det} g'_* \omega^{\otimes m}_{V'_{U^\dagger}/W'_{U^\dagger}} \hookrightarrow (\beta \circ \tau)^* \det (f_* \omega^{\otimes m}_{X_{U^\dagger}/U^\dagger}).
$$

Therefore, we obtain an injection

$$
\widehat{\det} g'_* \omega^{\otimes m}_{V'/W'} \hookrightarrow (\beta \circ \tau)^* \widehat{\det} (f_* \omega^{\otimes m}_{X/Y}) \otimes O_{W'}(E^\dagger)
$$

for some effective $(\beta \circ \tau)$-exceptional divisor $E^\dagger$ on W'. Thus, we obtain

$$
\begin{aligned}
\kappa(Y, \widehat{\det}(f_* \omega^{\otimes m}_{X/Y})) &= \kappa(W', (\beta \circ \tau)^* \widehat{\det}(f_* \omega^{\otimes m}_{X/Y}) \otimes \mathcal{O}_{W'}(E^\dagger)) \\
&\geq \kappa(W', \widehat{\det} g'_* \omega^{\otimes m}_{V'/W'}) \\
&= \dim W'' = \mathrm{Var}(f).
\end{aligned}
$$

This is the desired inequality. $\square$

For some future references, we write the following lemmas. The proof of Lemma 4.1.8 says:

Lemma 4.1.9 *Let $f : X \to Y$ be a surjective morphism between smooth projective varieties and let $\tau : Y' \to Y$ be a generically finite surjective morphism from a smooth projective variety Y'. We take the following commutative diagram:*

$$
\begin{array}{ccc}
X & \longleftarrow & X' \\
{\scriptstyle f}\downarrow & & \downarrow{\scriptstyle f'} \\
Y & \xleftarrow{\ \tau\ } & Y'
\end{array}
$$

where X' is a resolution of the main component of $X \times_Y Y'$. Let m be a positive integer. Then there exists an effective τ-exceptional divisor E on Y' such that

$$\widehat{\det} f'_* \omega_{X'/Y'}^{\otimes m} \hookrightarrow \tau^*(\widehat{\det} f_* \omega_{X/Y}^{\otimes m}) \otimes O_{Y'}(E).$$

In particular, we have

$$\kappa(Y, \widehat{\det} f_* \omega_{X/Y}^{\otimes m}) \geq \kappa(Y', \widehat{\det} f'_* \omega_{X'/Y'}^{\otimes m}).$$

Lemma 4.1.10 *Let $f : X \to Y$ be a surjective morphism between smooth projective varieties and let $\tau : Y' \to Y$ be a generically finite surjective morphism from a smooth projective variety Y'. We take the following commutative diagram:*

$$
\begin{array}{ccc}
X & \longleftarrow & X' \\
\downarrow f & & \downarrow f' \\
Y & \xleftarrow{\ \tau\ } & Y'
\end{array}
$$

where X' is a resolution of the main component of $X \times_Y Y'$. Let k be a positive integer. If $f'_* \omega_{X'/Y'}^{\otimes k}$ is big, then $f_* \omega_{X/Y}^{\otimes k}$ is also big.

Proof Let $\mathcal{H}$ be an ample invertible sheaf on Y. Since $f'_* \omega_{X'/Y'}^{\otimes k}$ is big by assumption, we have a positive integer a such that $\widehat{S}^a(f'_* \omega_{X'/Y'}^{\otimes k}) \otimes \tau^* \mathcal{H}^{\otimes -1}$ is weakly positive. By replacing Y with $Y \setminus \Sigma$ for some closed subset Σ of codimension ≥ 2 (see Lemma 3.1.12(i)), we may assume that τ is flat, $f_* \omega_{X/Y}^{\otimes k}$ and $f'_* \omega_{X'/Y'}^{\otimes k}$ are both locally free. By Lemma 3.1.20(i), we have the following generically isomorphic inclusion:

$$f'_* \omega_{X'/Y'}^{\otimes k} \subset \tau^*(f_* \omega_{X/Y}^{\otimes k}).$$

Therefore, we obtain

$$S^a(f'_* \omega_{X'/Y'}^{\otimes k}) \otimes \tau^* \mathcal{H}^{\otimes -1} \subset \tau^*(S^a(f_* \omega_{X/Y}^{\otimes k}) \otimes \mathcal{H}^{\otimes -1}),$$

which is generically an isomorphism. By Lemma 3.1.12(ii), $\tau^*(S^a(f_* \omega_{X/Y}^{\otimes k}) \otimes \mathcal{H}^{\otimes -1})$ is weakly positive since $S^a(f'_* \omega_{X'/Y'}^{\otimes k}) \otimes \tau^* \mathcal{H}^{\otimes -1}$ is weakly positive. By Lemma 3.1.12(v), $S^a(f_* \omega_{X/Y}^{\otimes k}) \otimes \mathcal{H}^{\otimes -1}$ is weakly positive since τ is finite. This means that $f_* \omega_{X/Y}^{\otimes k}$ is big. $\qquad\square$

We close this section with a useful remark.

Remark 4.1.11 As in 3.1.26, by Lemmas 4.1.9 and 4.1.10, we may assume that f is semistable in codimension one or

$$f : X \xrightarrow{\ \delta\ } X^\dagger \xrightarrow{\ f^\dagger\ } Y$$

such that $f^\dagger : X^\dagger \to Y$ is weakly semistable and that δ is a resolution of singularities when we want to prove that $\kappa(Y, \widehat{\det}(f_* \omega_{X/Y}^{\otimes m})) = \dim Y$ holds or that $f_* \omega_{X/Y}^{\otimes k}$ is big.

4.2 Fiber Spaces Whose Base Space is of General Type

In this short section, we prove the following theorem due to Viehweg as an easy application of Viehweg's weak positivity of direct images of relative pluricanonical bundles (see Theorems 3.2.5 and 3.3.5).

Theorem 4.2.1 ([Vi3, Corollary IV]) *Let $f : X \to Y$ be a surjective morphism between smooth projective varieties with connected fibers. Assume that Y is of general type, that is, $\kappa(Y) = \dim Y$. Then we have*

$$\kappa(X) = \kappa(X_y) + \kappa(Y)$$
$$= \kappa(X_y) + \dim Y,$$

where X_y is a sufficiently general fiber of $f : X \to Y$.

Let us start with the following easy lemma, which is an application of the flattening theorem (see Step 1 in the proof of Lemma 4.1.5).

Lemma 4.2.2 ([Vi3, Lemma 7.3]) *Let $f : X \to Y$ be a surjective morphism between smooth projective varieties with connected fibers. Then we can construct a commutative diagram:*

$$\begin{array}{ccc} X' & \xrightarrow{\ \rho\ } & X \\ {\scriptstyle f'}\downarrow & & \downarrow{\scriptstyle f} \\ Y' & \xrightarrow{\ \tau\ } & Y \end{array}$$

satisfying the following properties:

(i) $\tau : Y' \to Y$ is a birational morphism between smooth projective varieties.
(ii) $\rho : X' \to X$ is a birational morphism between smooth projective varieties.
(iii) Let D be any f'-exceptional prime divisor on X'. Then D is ρ-exceptional.

Proof By the flattening theorem (see, for example, [AbO, 3.3. The flattening lemma]), there is a birational morphism $\tau : Y' \to Y$ from a smooth projective variety Y' such that the main component of $X \times_Y Y'$ is flat over Y'. By taking a resolution of the main component of $X \times_Y Y'$, we obtain $f' : X' \to Y'$ with the desired properties.   □

Proof of Theorem 4.2.1 We note that the inequality $\kappa(X) \leq \kappa(X_y) + \dim Y$ always holds by the easy addition formula (see Lemma 2.3.31). If $\kappa(X_y) = -\infty$, then the desired equality holds true since we have $\kappa(X) = -\infty$ by the easy addition formula (see Lemma 2.3.31). Therefore, it is sufficient to prove

$$\kappa(X) \geq \kappa(X_y) + \kappa(Y) = \kappa(X_y) + \dim Y$$

under the assumption that $\kappa(X_y) \geq 0$. We construct a commutative diagram:

$$
\begin{array}{ccc}
X' & \xrightarrow{\ \rho\ } & X \\
{\scriptstyle f'}\downarrow & & \downarrow{\scriptstyle f} \\
Y' & \xrightarrow{\ \tau\ } & Y
\end{array}
$$

as in Lemma 4.2.2. We see that $f_*\omega_{X/Y}^{\otimes k}$ is nontrivial for some positive integer k since $\kappa(X_y) \geq 0$. Therefore, by Theorem 3.3.5, $f'_*\omega_{X'/Y'}^{\otimes l}$ is nontrivial and weakly positive for every positive integer l which is divisible by k. We take an ample invertible sheaf $\mathcal{H}'$ on Y'. By Kodaira's lemma (see Lemma 2.3.27), we obtain an inclusion $\mathcal{H}' \hookrightarrow \omega_{Y'}^{\otimes l}$ if l is a sufficiently large positive integer. This is because $\omega_{Y'}$ is big, that is, $\kappa(Y', \omega_{Y'}) = \kappa(Y') = \kappa(Y) = \dim Y$. Without loss of generality, we may further assume that l is divisible by k. Since $f'_*\omega_{X'/Y'}^{\otimes l}$ is weakly positive, there is a positive integer β such that

$$
\widehat{S}^{2\beta}(f'_*\omega_{X'/Y'}^{\otimes l}) \otimes \mathcal{H}'^{\otimes \beta}
$$

is generically generated by global sections. By Remark 3.1.2, that is, by replacing β with a multiple, we may assume that $\mathcal{H}'^{\otimes \beta}$ is very ample. We take an effective f'-exceptional divisor D on X' such that

$$
\widehat{S}^1(f'_*\omega_{X'/Y'}^{\otimes 2l\beta}) = \left(f'_*\omega_{X'/Y'}^{\otimes 2l\beta}\right)^{**} = f'_*\left(\omega_{X'/Y'}^{\otimes 2l\beta} \otimes \mathcal{O}_{X'}(2l\beta D)\right)
$$

holds. Then the natural map

$$
\widehat{S}^{2\beta}\left(f'_*\omega_{X'/Y'}^{\otimes l}\right) \otimes \mathcal{H}'^{\otimes \beta} \to f'_*\left(\omega_{X'/Y'}^{\otimes 2l\beta} \otimes \mathcal{O}_{X'}(2l\beta D)\right) \otimes \mathcal{H}'^{\otimes \beta}
$$

is nontrivial. Therefore, we obtain an inclusion of $f'^*\mathcal{H}'^{\otimes \beta}$ into

$$
\left(\omega_{X'/Y'}^{\otimes 2l\beta} \otimes \mathcal{O}_{X'}(2l\beta D)\right) \otimes f'^*\mathcal{H}'^{\otimes 2\beta}.
$$

By Lemma 4.1.4, we obtain

$$
\kappa(X', \omega_{X'/Y'}^{\otimes l} \otimes \mathcal{O}_{X'}(lD) \otimes f'^*\mathcal{H}') = \kappa(X'_y) + \dim Y',
$$

where X'_y is a sufficiently general fiber of $f' : X' \to Y'$. Therefore, we have

$$
\begin{aligned}
\kappa(X) &= \kappa(X') \\
&= \kappa(X', \omega_{X'} \otimes \mathcal{O}_{X'}(D)) \\
&\geq \kappa(X', \omega_{X'/Y'}^{\otimes l} \otimes \mathcal{O}_{X'}(lD) \otimes f'^*\mathcal{H}') \\
&= \kappa(X'_y) + \dim Y' \\
&= \kappa(X_y) + \kappa(Y).
\end{aligned}
$$

The first equality is obvious since X and X' are smooth projective varieties and they are birationally equivalent. The second equality holds since D is ρ-exceptional by Lemma 4.2.2(iii). The third inequality follows from the inclusion $\mathcal{H}' \hookrightarrow \omega_{Y'}^{\otimes l}$. The fourth equality is a consequence of Fujita's lemma (see Lemma 4.1.4) explained above. The final equality is obvious since $\kappa(X'_y) = \kappa(X_y)$ and $\dim Y' = \dim Y = \kappa(Y)$.   □

We close this short section with a remark on a logarithmic generalization of Theorem 4.2.1.

Remark 4.2.3 Theorem 4.2.1 can be generalized for the logarithmic Kodaira dimension. Precisely speaking, Conjecture 1.2.2 holds true under the assumption that $\overline{\kappa}(Y) = \dim Y$ (see [Mae, Corollary 2]). For the details, see [F15, Sect. 10].

4.3 Fiber Spaces Whose Geometric Generic Fiber is of General Type

In this section, we discuss projective surjective morphisms between smooth projective varieties whose geometric generic fiber is of general type. The main purpose of this section is to prove:

Theorem 4.3.1 *Let $f : X \to Y$ be a surjective morphism between smooth projective varieties with connected fibers. Assume that the geometric generic fiber $X_{\overline{\eta}}$ of $f : X \to Y$ is of general type. Then there exists a generically finite surjective morphism $\tau : Y' \to Y$ from a smooth projective variety Y' with the following property.*

Let X' be any resolution of the main component of $X \times_Y Y'$ sitting in the commutative diagram below:

$$
\begin{array}{ccc}
X' & \longrightarrow & X \\
{\scriptstyle f'}\downarrow & & \downarrow{\scriptstyle f} \\
Y' & \xrightarrow{\ \tau\ } & Y.
\end{array}
$$

Then $f'_* \omega_{X'/Y'}^{\otimes m}$ is a nef locally free sheaf for every nonnegative integer m. In particular, $\det f'_* \omega_{X'/Y'}^{\otimes m}$ is a nef invertible sheaf for every nonnegative integer m. We further assume that $\mathrm{Var}(f) = \dim Y$. Then $\det f'_* \omega_{X'/Y'}^{\otimes k}$ is a nef and big invertible sheaf for some large and divisible positive integer k.

Theorem 4.3.1 is better than the well-known results by Kawamata, Kollár, Viehweg, and others (see [Kaw6], [Kol3], and [Vi6]).

We have more informations on $\tau : Y' \to Y$ by the proof of Theorem 4.3.1 below. The following remark is very important for various applications.

Remark 4.3.2 In Theorem 4.3.1, let $\tau : Y' \to Y$ be a generically finite surjective morphism from a smooth projective variety Y' such that there exists a weakly semistable morphism $f^\dagger : X^\dagger \to Y'$ in the sense of Abramovich–Karu (see 2.3.42), where $X^\dagger$ is birationally equivalent to the main component of $X \times_Y Y'$ over Y'. Then the proof of Theorem 4.3.1 says that $f'_* \omega_{X'/Y'}^\otimes$ is a nef locally free sheaf for every nonnegative integer m and that $\det f'_* \omega_{X'/Y'}^{\otimes k}$ is a nef and big invertible sheaf for some large and divisible positive integer k when $\mathrm{Var}(f) = \dim Y$.

Remark 4.3.3 In Theorem 4.3.1, if $f' : X' \to Y'$ satisfies some mild assumptions, then the bigness of $\det f'_* \omega_{X'/Y'}^{\otimes k}$ implies that $f'_* \omega_{X'/Y'}^{\otimes k'}$ is a big locally free sheaf, where k' is any multiple of k with $k' \geq 2$. For the details, see Theorem 3.3.12 and Remark 3.3.13.

Therefore, we have:

Theorem 4.3.4 *Let $f : X \to Y$ be a surjective morphism between smooth projective varieties with connected fibers such that the geometric generic fiber $X_{\overline{\eta}}$ of $f : X \to Y$ is of general type. Assume that $\mathrm{Var}(f) = \dim Y$. Then $f_* \omega_{X/Y}^{\otimes k}$ is big for some positive integer k.*

Theorem 4.3.4 says that Viehweg's Conjecture Q (see Conjecture 1.2.6) holds true under the assumption that the geometric generic fiber is of general type.

Proof of Theorem 4.3.4 By the weak semistable reduction theorem (see Theorem 2.3.46) and Lemma 4.1.10, we may assume that $f : X \to Y$ decomposes as

$$f : X \xrightarrow{\delta} X^\dagger \xrightarrow{f^\dagger} Y$$

such that $f^\dagger$ is weakly semistable and that δ is a resolution of singularities (see Remark 4.1.11). Then, by Theorem 4.3.1 and Remark 4.3.2, $\det f_* \omega_{X/Y}^{\otimes k_0}$ is nef and big for some positive integer k_0. By Theorem 3.3.12 and Remark 3.3.13, $f_* \omega_{X/Y}^{\otimes k}$ is big for some positive integer k. $\qquad\square$

By the results in Sect. 4.1, we have the following result as an application of Theorem 4.3.1 (and Remark 4.3.2). Corollary 4.3.5 says that the generalized Iitaka conjecture (see Conjecture 1.2.5) holds for projective surjective morphisms between smooth projective varieties with connected fibers whose geometric generic fiber is of general type.

Corollary 4.3.5 (see [Kaw6], [Kol3], and [Vi6]) *Let $f : X \to Y$ be a surjective morphism between smooth projective varieties with connected fibers. Assume that the geometric generic fiber $X_{\overline{\eta}}$ of $f : X \to Y$ is of general type. Then we have the following properties:*

(i) There exists a positive integer k such that

$$\kappa(Y, \widehat{\det}(f_* \omega_{X/Y}^{\otimes k})) \geq \mathrm{Var}(f).$$

(ii) *Let $\mathcal{L}$ be an invertible sheaf on Y. If $\kappa(Y, \mathcal{L}) \geq 0$, then we have*

$$\kappa(X, \omega_{X/Y} \otimes f^*\mathcal{L}) \geq \kappa(X_{\overline{\eta}}) + \max\{\kappa(Y, \mathcal{L}), \mathrm{Var}(f)\}$$
$$= \dim X - \dim Y + \max\{\kappa(Y, \mathcal{L}), \mathrm{Var}(f)\}.$$

(iii) We have

$$\kappa(X, \omega_{X/Y}) \geq \kappa(X_{\overline{\eta}}) + \mathrm{Var}(f)$$
$$= \dim X - \dim Y + \mathrm{Var}(f).$$

(iv) If $\kappa(Y) \geq 0$, then we have

$$\kappa(X) \geq \kappa(X_{\overline{\eta}}) + \max\{\kappa(Y), \mathrm{Var}(f)\}$$
$$= \dim X - \dim Y + \max\{\kappa(Y), \mathrm{Var}(f)\}.$$

Proof Note that (iii) and (iv) are important special cases of the statement (ii). In the big commutative diagram constructed in Lemma 4.1.5, we apply Theorem 4.3.1 to $g'' : V'' \to W''$ (see also Remark 4.3.2). Then we obtain that $\det g''_* \omega^{\otimes m}_{V''/W''}$ is nef and big for some positive integer m. Therefore, we obtain the desired inequality in (ii) by Lemma 4.1.7. We also obtain the desired inequality in (i) by Lemma 4.1.8. $\qquad\square$

Before we start the proof of Theorem 4.3.1, we prepare several lemmas for the reader's convenience.

Lemma 4.3.6 *Let X be a normal variety with only canonical singularities. Then $\mathcal{O}_X(mK_X)$ is Cohen–Macaulay for every integer m.*

Proof We note that X has only rational singularities when X is canonical. Let r be the smallest positive integer such that rK_X is Cartier. Since the problem is local, we may assume that $rK_X \sim 0$ by shrinking X. If $r = 1$, then $\mathcal{O}_X(mK_X) \simeq \mathcal{O}_X$ for every integer m. In this case, $\mathcal{O}_X(mK_X)$ is Cohen–Macaulay for every integer m since X has only rational singularities. From now on, we assume that $r \geq 2$. Let $\pi : \widetilde{X} \to X$ be the index one cover. Then we have

$$\pi_* \mathcal{O}_{\widetilde{X}}(K_{\widetilde{X}}) \simeq \bigoplus_{i=1}^{r} \mathcal{O}_X(iK_X).$$

Since $\widetilde{X}$ has only canonical singularities and $K_{\widetilde{X}}$ is Cartier, $\mathcal{O}_{\widetilde{X}}(K_{\widetilde{X}})$ is Cohen–Macaulay. Since π is finite, $\mathcal{O}_X(iK_X)$ is Cohen–Macaulay for $1 \leq i \leq r$. By $rK_X \sim 0$, we obtain that $\mathcal{O}_X(mK_X)$ is Cohen–Macaulay for every integer m. $\qquad\square$

Let us recall the following well-known lemma, which is a special case of [N1, Corollary 3].

Lemma 4.3.7 (cf. [N1, Corollary 3]) *Let $g : V \to C$ be a projective surjective morphism from a normal quasi-projective variety V to a smooth quasi-projective curve*

C. Assume that V has only canonical singularities and that K_V is g-semi-ample. Then $R^i g_* \mathcal{O}_V(mK_V)$ is locally free for every i and every positive integer m.

Proof Let $h : V' \to V$ be a resolution of singularities such that $\text{Exc}(h)$ is a simple normal crossing divisor on V'. We write

$$K_{V'} = h^* K_V + E,$$

where E is an effective h-exceptional $\mathbb{Q}$-divisor. Then we have

$$\lceil mh^* K_V + E \rceil - (K_{V'} + \{-(mh^* K_V + E)\}) = (m-1)h^* K_V.$$

We note that the righthand side is semi-ample over C. Therefore,

$$R^i(g \circ h)_* \mathcal{O}_{V'}(\lceil mh^* K_V + E \rceil)$$

is locally free for every i and every positive integer m (see, for example, Theorem 2.3.50(i) and [F7, Theorem 6.3(i)]). On the other hand, we have

$$R^i h_* \mathcal{O}_{V'}(\lceil mh^* K_V + E \rceil) = 0$$

for every $i > 0$ by the relative Kawamata–Viehweg vanishing theorem, and

$$h_* \mathcal{O}_{V'}(\lceil mh^* K_V + E \rceil) \simeq \mathcal{O}_V(mK_V).$$

Therefore, we obtain that

$$R^i g_* \mathcal{O}_V(mK_V)$$

is locally free for every i and every positive integer m. $\qquad\square$

We will use the following easy criterion of nefness in the proof of Theorem 4.3.1.

Lemma 4.3.8 *Let $\mathcal{E}$ be a nonzero locally free sheaf of finite rank on a smooth projective variety V. Assume that there exists an invertible sheaf $\mathcal{M}$ such that $\mathcal{E}^{\otimes s} \otimes \mathcal{M}$ is generated by global sections for every positive integer s. Then $\mathcal{E}$ is nef.*

Proof We put $\pi : W = \mathbb{P}_V(\mathcal{E}) \to V$ and $\mathcal{O}_W(1) = \mathcal{O}_{\mathbb{P}_V(\mathcal{E})}(1)$. Since $\mathcal{E}^{\otimes s} \otimes \mathcal{M}$ is generated by global sections, $S^s(\mathcal{E}) \otimes \mathcal{M}$ is also generated by global sections for every positive integer s. This implies that $\mathcal{O}_W(s) \otimes \pi^* \mathcal{M}$ is generated by global sections for every positive integer s. Thus, we obtain that $\mathcal{O}_W(1)$ is nef, equivalently, $\mathcal{E}$ is nef. $\qquad\square$

Let us start the proof of Theorem 4.3.1.

Proof of Theorem 4.3.1 Let us divide the proof into several steps. First, let us prove the existence of $f' : X' \to Y'$ such that $f'_* \omega_{X'/Y'}^{\otimes m}$ is locally free for every positive integer m.

Step 1 (Weak semistable reduction) By Theorem 2.3.46 (see also [AbK, Theorem 0.3]), there exist a generically finite morphism $\tau : Y' \to Y$ from a smooth projective variety Y' and $f^\dagger : X^\dagger \to Y'$ with the following properties:

(i) $X^\dagger$ is a normal projective Gorenstein (see Lemma 2.3.43 and [AbK, Lemma 6.1]) variety which is birationally equivalent to $X \times_Y Y'$.
(ii) $(U_{X^\dagger} \subset X^\dagger)$ and $(U_{Y'} \subset Y')$ are toroidal embeddings without self-intersection, with $U_{X^\dagger} = (f^\dagger)^{-1}(U_{Y'})$.
(iii) $f^\dagger : (U_{X^\dagger} \subset X^\dagger) \to (U_{Y'} \subset Y')$ is toroidal and equidimensional.
(iv) All the fibers of the morphism $f^\dagger$ are reduced.

Note that $f^\dagger : X^\dagger \to Y'$ is weakly semistable (see 2.3.42) and is called a weak semistable reduction of $f : X \to Y$. We also note that $X^\dagger$ has only rational singularities since $X^\dagger$ is toroidal. Therefore, $X^\dagger$ has only canonical Gorenstein singularities (see Lemma 2.3.13) and is Cohen–Macaulay. Thus, we have

$$f^\dagger_* \mathcal{O}_{X^\dagger}(mK_{X^\dagger/Y'}) \simeq f'_* \omega^{\otimes m}_{X'/Y'}$$

for every positive integer m. We note that $K_{X^\dagger/Y'} = K_{X^\dagger} - f^{\dagger *} K_{Y'}$. Therefore, it is sufficient to prove that $f^\dagger_* \mathcal{O}_{X^\dagger}(mK_{X^\dagger/Y'})$ is locally free for every positive integer m. Note that $f^\dagger$ is flat since Y' is smooth, $X^\dagger$ is Cohen–Macaulay, and $f^\dagger$ is equidimensional (see [Har3, Chap. III, Exercise 10.9] and [AlK, Chap. V, Proposition (3.5)]).

Step 2 (Relative canonical models) By assumption, the geometric generic fiber of $f^\dagger : X^\dagger \to Y'$ is of general type. Therefore, $f^\dagger : X^\dagger \to Y'$ has the relative canonical model $\widetilde{f} : \widetilde{X} \to Y'$ by [BCHM], that is, there exists a commutative diagram:

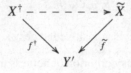

such that:

(i) $X^\dagger \dashrightarrow \widetilde{X}$ is birational,
(ii) $\widetilde{f} : \widetilde{X} \to Y'$ is projective,
(iii) $K_{\widetilde{X}}$ is $\widetilde{f}$-ample, and
(iv) $\widetilde{X}$ has only canonical singularities.

We note that the relative canonical model is unique and that $\widetilde{f} : \widetilde{X} \to Y'$ is nothing but

$$\mathrm{Proj}_{Y'} \bigoplus_{m \geq 0} f^\dagger_* \mathcal{O}_{X^\dagger}(mK_{X^\dagger/Y'}) \to Y'.$$

We also note that

$$f^\dagger_* \mathcal{O}_{X^\dagger}(mK_{X^\dagger/Y'}) \simeq \widetilde{f}_* \mathcal{O}_{\widetilde{X}}(mK_{\widetilde{X}/Y'})$$

for every positive integer m since both $X^\dagger$ and $\widetilde{X}$ have only canonical singularities. Therefore, it is sufficient to prove that $\widetilde{f}_*\mathcal{O}_{\widetilde{X}}(m K_{\widetilde{X}/Y'})$ is locally free for every positive integer m.

Step 3 (see [F11]) We take an arbitrary closed point $P \in Y'$. We take general very ample Cartier divisors $H_1, H_2, \cdots, H_{n-1}$, where $n = \dim Y$, such that $C = H_1 \cap H_2 \cap \cdots \cap H_{n-1}$ is a smooth projective curve passing through P. We put

$$S_i = H_1 \cap \cdots \cap H_i$$

for every $1 \leq i \leq n - 1$. Of course, $S_1 = H_1$ and $S_{n-1} = C$. Thus we have a sequence of Cartier divisors

$$Y' =: S_0 \supset S_1 \supset \cdots \supset S_{n-1} = C.$$

We put $X^\dagger_{S_i} = X^\dagger \times_{Y'} S_i$ for every i. Then, by adjunction, $X^\dagger_{S_i}$ is Gorenstein for every i. Note that $X^\dagger$ is Gorenstein and that $X^\dagger_{S_{i+1}}$ is a Cartier divisor on $X^\dagger_{S_i}$ for every i. By Lemma 2.3.49 (see also [AbK, Lemma 6.2]), $X^\dagger_{S_i} \to S_i$ is weakly semistable outside P for every $1 \leq i \leq n - 1$. In particular, $X^\dagger_{S_i}$ is normal outside $(f^\dagger)^{-1}(P)$ for every $1 \leq i \leq n - 1$. Since $X^\dagger_{S_i}$ is Gorenstein and $\operatorname{codim}_{X^\dagger_{S_i}}(f^\dagger)^{-1}(P) \geq 2$ for every $1 \leq i \leq n - 2$, $X^\dagger_{S_i}$ is normal for every $1 \leq i \leq n - 2$. Note that $f^\dagger : X^\dagger \to Y'$ is equidimensional. By Lemma 2.3.49 (see also [Kar2, Lemma 2.12 (2)]), $X^\dagger_{S_{n-1}} \to S_{n-1} = C$ is weakly semistable because $\operatorname{Supp}(Y' \setminus U_{Y'})|_C = P$ in a neighborhood of P. Therefore, $X^\dagger_{S_{n-1}}$ is normal. Thus, $X^\dagger_{S_i}$ is normal for every i. We note that $X^\dagger_{S_i}$ has only rational Gorenstein singularities outside $(f^\dagger)^{-1}(P)$ for every i because $X^\dagger_{S_i} \to S_i$ is weakly semistable outside P for every i. We consider the pair $(X^\dagger_{S_{n-2}}, X^\dagger_{S_{n-1}})$. Note that $X^\dagger_{S_{n-1}}$ has only rational Gorenstein singularities because $X^\dagger_{S_{n-1}} \to S_{n-1}$ is weakly semistable (see Lemma 2.3.43 and [AbK, Lemma 6.1]). We also note that a normal variety has only rational Gorenstein singularities if and only if it has only canonical Gorenstein singularities (see Lemma 2.3.13). By the inversion of adjunction (see [KolM, Theorem 5.50]), $(X^\dagger_{S_{n-2}}, X^\dagger_{S_{n-1}})$ is plt in a neighborhood of $X^\dagger_{S_{n-1}}$. Therefore, $X^\dagger_{S_{n-2}}$ has only canonical Gorenstein singularities. Apply the same argument to the pair $(X^\dagger_{S_{i-1}}, X^\dagger_{S_i})$ for $n - 2 \geq i \geq 2$ inductively. Then $X^\dagger_{S_i}$ has only canonical Gorenstein singularities for every i. We put $\widetilde{X}_{S_i} = \widetilde{X} \times_{Y'} S_i$ for every i. Since $(X^\dagger, X^\dagger_{S_1})$ is plt by the inversion of adjunction as above (see [KolM, Theorem 5.50]), $(\widetilde{X}, \widetilde{X}_{S_1})$ is also plt by the negativity lemma (see, for example, [KolM, Proposition 3.51]). Thus, $\widetilde{X}_{S_1}$ is normal (see [KolM, Proposition 5.51]). By the negativity lemma and adjunction again, $\widetilde{X}_{S_1}$ has only canonical singularities. By repeating this process $(n - 1)$-times, we obtain that $\widetilde{X}_C = \widetilde{X} \times_{Y'} C$ has only canonical singularities. Since $\widetilde{X}_C \to C$ is a dominant morphism from a normal irreducible variety $\widetilde{X}_C$ onto a smooth curve C, it is flat. In particular, $\widetilde{X}_C \to C$ is equidimensional. Therefore, $\widetilde{f} : \widetilde{X} \to Y'$ is equidimensional by the choice of C. Since $\widetilde{X}$ is Cohen–Macaulay and Y' is smooth, $\widetilde{f}$ is flat (see [Har3, Chap. III, Exercise 10.9] and [AlK, Chap. V, Proposition (3.5)]).

Moreover, $\mathcal{O}_{\widetilde{X}}(mK_{\widetilde{X}})$ is flat over Y' for every integer m since $\mathcal{O}_{\widetilde{X}}(mK_{\widetilde{X}})$ is Cohen–Macaulay by Lemma 4.3.6 and $\widetilde{f}$ is equidimensional (see [AlK, Chap. V, Proposition (3.5)]). We note that $\mathcal{O}_{\widetilde{X}}(mK_{\widetilde{X}/Y'})|_{\widetilde{X}_C} \simeq \mathcal{O}_{\widetilde{X}_C}(mK_{\widetilde{X}_C/C})$ holds by repeatedly using adjunction. By applying Lemma 4.3.7 and the base change theorem (see [Har3, Chap. III, Theorem 12.11]) to $\widetilde{X}_C \to C$, we obtain that

$$\dim H^0(\widetilde{X}_y, \mathcal{O}_{\widetilde{X}}(mK_{\widetilde{X}/Y'})|_{\widetilde{X}_y})$$

is independent of $y \in Y'$ for every positive integer m. By the base change theorem (see [Har3, Chap. III, Corollary 12.9]), we obtain that $f'_*\omega_{X'/Y'}^{\otimes m} \simeq \widetilde{f}_*\mathcal{O}_{\widetilde{X}}(mK_{\widetilde{X}/Y'})$ is locally free for every positive integer m.

We complete the proof of the local freeness of $f'_*\omega_{X'/Y'}^{\otimes m}$.

Let us see several remarks on the above arguments, although we do not need them in the proof of Theorem 4.3.1.

Remark 4.3.9 If $m \geq 2$, then we obtain

$$R^i\widetilde{f}_*\mathcal{O}_{\widetilde{X}}(mK_{\widetilde{X}}) = 0$$

for every $i > 0$ by the relative Kawamata–Viehweg vanishing theorem (see, for example, [F12, Corollary 5.7.7]). We note that $K_{\widetilde{X}}$ is $\widetilde{f}$-ample by construction. Therefore, we see that $\widetilde{f}_*\mathcal{O}_{\widetilde{X}}(mK_{\widetilde{X}/Y'})$ is locally free for every $m \geq 2$. Unfortunately, this argument does not work for the case when $m = 1$.

Remark 4.3.10 By the argument in Step 3, Remark 4.3.9, Lemma 4.3.7, and the base change theorem, we obtain that $R^i\widetilde{f}_*\mathcal{O}_{\widetilde{X}}(mK_{\widetilde{X}/Y'}) = 0$ for every $i > 0$ and every $m \geq 2$, and that $R^i\widetilde{f}_*\mathcal{O}_{\widetilde{X}}(K_{\widetilde{X}/Y'})$ is locally free for every $i > 0$.

Remark 4.3.11 *(Adjunction)* In general, $\widetilde{X}_y$ may be nonnormal. However, we see that the canonical divisor $K_{\widetilde{X}_y}$ is well-defined, $\widetilde{X}_y$ has only semi-log-canonical singularities, and $\mathcal{O}_{\widetilde{X}}(mK_{\widetilde{X}/Y'})|_{\widetilde{X}_y} \simeq \mathcal{O}_{\widetilde{X}_y}(mK_{\widetilde{X}_y})$ for every positive integer m, by adjunction. For the details of semi-log-canonical singularities and pairs, see [F8].

Remark 4.3.12 Let $f : X \to Y$ be a surjective morphism with connected fibers from a normal projective variety X to a smooth projective variety Y. Assume that f is weakly semistable and the geometric generic fiber of f has a good minimal model. Then we have already proved that $f_*\omega_{X/Y}^{\otimes m}$ is locally free for every positive integer m. For the details, see [F10] and [F11] (see also [F14, Sect. 3]).

Next, we will prove that $f'_*\omega_{X'/Y'}^{\otimes m}$ is nef. Our proof depends on the effective freeness due to Popa–Schnell (see Theorem 3.2.1). We do not need the difficult semipositivity theorem in [F16] (see also [F17]).

Step 4 (Nefness) By the proof of the local freeness of $f'_*\omega_{X'/Y'}^{\otimes m}$, we may assume that $f' : X' \to Y'$ is weakly semistable. For simplicity of notation, we denote $f' : X' \to Y'$ by $f : X \to Y$ in this step. We take the s-fold fiber product

$$f^s : X^s = X \times_Y X \times_Y \cdots \times_Y X \to Y.$$

Then we see that X^s is normal and Gorenstein. Moreover, X^s has only rational singularities because X^s is local analytically isomorphic to a toric variety. Therefore, X^s has only canonical singularities. For the details, see Lemma 2.3.48 and Remark 3.3.13. We note the following commutative diagram:

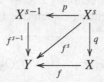

By the flat base change theorem [Ve, Theorem 2] (see also [Har2], [Co], and so on), we have $\omega_{X^s/X} \simeq p^* \omega_{X^{s-1}/Y}$. Thus we have

$$\omega_{X^s/Y} \simeq \omega_{X^s/X} \otimes q^* \omega_{X/Y}$$
$$\simeq p^* \omega_{X^{s-1}/Y} \otimes q^* \omega_{X/Y}.$$

Therefore, by the flat base change theorem (see [Har3, Chap. III, Proposition 9.3]) and the projection formula, we obtain

$$f^s_* \omega^{\otimes m}_{X^s/Y} \simeq f^{s-1}_* p_*(p^* \omega^{\otimes m}_{X^{s-1}/Y} \otimes q^* \omega^{\otimes m}_{X/Y})$$
$$\simeq f^{s-1}_*(\omega^{\otimes m}_{X^{s-1}/Y} \otimes p_* q^* \omega^{\otimes m}_{X/Y})$$
$$\simeq f^{s-1}_*(\omega^{\otimes m}_{X^{s-1}/Y} \otimes (f^{s-1})^* f_* \omega^{\otimes m}_{X/Y})$$
$$\simeq f_* \omega^{\otimes m}_{X/Y} \otimes f^{s-1}_* \omega^{\otimes m}_{X^{s-1}/Y}$$
$$\simeq \bigotimes^s f_* \omega^{\otimes m}_{X/Y}$$

for every positive integer m and every positive integer s by induction on s. Note that $f_* \omega^{\otimes m}_{X/Y}$ is locally free for every positive integer m. By Corollary 3.2.7, we see that

$$f^s_* \omega^{\otimes m}_{X^s/Y} \otimes \omega^{\otimes m}_Y \otimes \mathcal{L}^{\otimes m(\dim Y+1)}$$
$$\simeq \left(\bigotimes^s f_* \omega^{\otimes m}_{X/Y} \right) \otimes \omega^{\otimes m}_Y \otimes \mathcal{L}^{\otimes m(\dim Y+1)}$$

is generated by global sections for every positive integer s, where $\mathcal{L}$ is an ample invertible sheaf on Y such that $|\mathcal{L}|$ is free. Therefore, by Lemma 4.3.8, we obtain that the locally free sheaf $f_* \omega^{\otimes m}_{X/Y}$ is nef for every positive integer m.

Remark 4.3.13 Let $f : X \to Y$ be a surjective morphism from a normal projective variety X to a smooth projective variety Y. Let m be a positive integer. The above argument shows that $f_*\omega_{X/Y}^{\otimes m}$ is nef if f is weakly semistable and $f_*\omega_{X/Y}^{\otimes m}$ is locally free.

Finally, we will prove that $\det f'_*\omega_{X'/Y'}^{\otimes k}$ is big for some positive integer k under the assumption that $\mathrm{Var}(f) = \dim Y$. We closely follow the proof of [Vi7, Theorem 4.34] (see also [Kol4, 3.13. Lemma]).

Step 5 (Bigness) In this step, we denote $\widetilde{f} : \widetilde{X} \to Y'$ by $f : X \to Y$ for simplicity of notation. We take a positive integer l such that $lK_{X/Y}$ is f-very ample and that the multiplication map

$$\delta : S^\mu(f_*\mathcal{O}_X(lK_{X/Y})) \to f_*\mathcal{O}_X(\mu l K_{X/Y}) \tag{4.3.1}$$

is surjective for every positive integer μ. We put $\mathcal{E} = f_*\mathcal{O}_X(lK_{X/Y})$. Then we obtain the following commutative diagram:

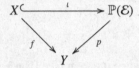

If $\mathcal{I}$ is the ideal sheaf of $\iota(X)$ on $\mathbb{P}(\mathcal{E})$, then we can find some positive integer μ such that

$$p^*p_*(\mathcal{I} \otimes \mathcal{O}_{\mathbb{P}(\mathcal{E})}(\mu)) \to \mathcal{I} \otimes \mathcal{O}_{\mathbb{P}(\mathcal{E})}(\mu) \tag{4.3.2}$$

is surjective. We fix this positive integer μ throughout this step.

Remark 4.3.14 The map δ in (4.3.1) is nothing but the restriction map

$$p_*\mathcal{O}_{\mathbb{P}(\mathcal{E})}(\mu) \to f_*(\mathcal{O}_{\mathbb{P}(\mathcal{E})}(\mu)|_X).$$

Therefore, we obtain

$$\mathrm{Ker}\delta = p_*(\mathcal{I} \otimes \mathcal{O}_{\mathbb{P}(\mathcal{E})}(\mu)).$$

By (4.3.2), $\mathrm{Ker}\delta$ recovers $\mathcal{I}$. Thus, $\mathrm{Ker}\delta$ recovers $f : X \to Y$.

Let us continue the proof. We consider

$$\mathbb{P} = \mathbb{P}(\bigoplus^r \mathcal{E}^*) \xrightarrow{\pi} Y$$

for $r = \mathrm{rank}\mathcal{E}$. We have the universal basis map

$$s : \bigoplus^r \mathcal{O}_\mathbb{P}(-1) \to \pi^*\mathcal{E}.$$

The map s is injective. Let Δ be the zero divisor of $\det(s)$. We put $Q = f_* \mathcal{O}_X(\mu l K_{X/Y})$ and consider the surjective map

$$\delta : S^\mu(\mathcal{E}) \to Q.$$

Let $\mathcal{B} \subset \pi^*Q$ be the image of the morphism

$$S^\mu(\overset{r}{\bigoplus} \mathcal{O}_{\mathbb{P}}(-1)) = S^\mu(\overset{r}{\bigoplus} \mathcal{O}_{\mathbb{P}}) \otimes \mathcal{O}_{\mathbb{P}}(-\mu) \xrightarrow{S^\mu(s)} S^\mu(\pi^*\mathcal{E}) \xrightarrow{\pi^*(\delta)} \pi^*Q.$$

By taking blow-ups of $\mathbb{P}$ with centers in Δ, we can obtain a projective birational morphism $\tau : \mathbb{P}' \to \mathbb{P}$ such that $\mathcal{B}' = \tau^*\mathcal{B}/\text{torsion}$ is locally free. We put $\mathcal{O}_{\mathbb{P}'}(1) = \tau^*\mathcal{O}_{\mathbb{P}}(1)$ and $\pi' = \pi \circ \tau$. Then we obtain a surjective morphism

$$\theta : S^\mu(\overset{r}{\bigoplus} \mathcal{O}_{\mathbb{P}'}(-1)) \to \mathcal{B}'.$$

We have the Plücker embedding

$$\text{Grass}(\text{rank}(Q), S^\mu(\mathbb{C}^r)) \hookrightarrow \mathbb{P}^M$$

and the surjection θ corresponds to the morphism

$$\rho' : \mathbb{P}' \to \text{Grass}(\text{rank}(Q), S^\mu(\mathbb{C}^r)) \hookrightarrow \mathbb{P}^M$$

such that

$$\det(\mathcal{B}') \otimes \mathcal{O}_{\mathbb{P}'}(\gamma) \simeq \rho'^*\mathcal{O}_{\mathbb{P}^M}(1),$$

where $\gamma = \mu \cdot \text{rank}Q$. By assumption, we have $\text{Var}(f) = \dim Y$. Note that the general fiber X_y of $f : X \to Y$ is a canonically polarized variety with only canonical singularities. Thus, the automorphism group of X_y is finite (see, for example, [U1, Corollary 14.3]). More precisely, we consider the embedding

$$\Phi_{|lK_{X_y}|} : X_y \hookrightarrow \mathbb{P}^{r-1}.$$

Then

$$\#\{g \in \text{PGL}(r, \mathbb{C}) \mid g(X_y) = X_y\}$$

is finite. In our situation, we consider

$$X_y \overset{\Phi_{|lK_{X_y}|}}{\hookrightarrow} \mathbb{P}^{r-1} \overset{\nu_\mu}{\hookrightarrow} \mathbb{P}^d$$

with

$$d = \binom{\mu + r - 1}{r - 1} - 1,$$

where ν_μ is the Veronese embedding of degree μ. Then we can easily check that $\mathrm{PGL}(r, \mathbb{C})$ acts on $\mathbb{P}^d$ and

$$\#\{g \in \mathrm{PGL}(r, \mathbb{C}) \mid g(X_y) = X_y\}$$

is finite, that is, the number of the elements of $\mathrm{PGL}(r, \mathbb{C})$ whose action on $\mathbb{P}^d$ preserves X_y is finite. Therefore, $\rho' : \mathbb{P}'_y = (\pi \circ \tau)^{-1}(y) \to \rho'(\mathbb{P}'_y)$ is generically finite. We assume that $\rho' : \mathbb{P}' \to \rho'(\mathbb{P}')$ is not generically finite. We put $Y' = (\rho')^* H_1 \cap \cdots \cap (\rho')^* H_{r^2-1}$, where $H_1, \ldots, H_{r^2-1}$ are general hyperplanes of $\mathbb{P}^M$. Then Y' is a smooth subvariety of $\mathbb{P}'$ such that $Y' \to Y$ is generically finite and $\rho' : Y' \to \rho'(Y')$ is not generically finite. Let X' be a resolution of the main component of $X \times_Y Y'$ and let $f' : X' \to Y'$ be the induced morphism. Then we have

$$\dim Y = \mathrm{Var}(f) = \mathrm{Var}(f') < \dim Y' = \dim Y$$

by construction. This is a contradiction. Therefore, we obtain that the morphism $\rho' : \mathbb{P}' \to \mathbb{P}^M$ is generically finite over its image. Thus $\rho'^* \mathcal{O}_{\mathbb{P}^M}(1)$ is nef and big on $\mathbb{P}'$. Let H be an ample Cartier divisor on Y. By Kodaira's lemma (see Lemma 2.3.27), we have

$$H^0(\mathbb{P}', \rho'^* \mathcal{O}_{\mathbb{P}^M}(\nu) \otimes \pi'^* \mathcal{O}_Y(-H)) \neq 0$$

for some large positive integer ν. Note that $\pi'^* Q$ and its subsheaf $\mathcal{B}'$ coincide over a nonempty Zariski open set of $\mathbb{P}'$. Thus

$$\pi'^*(\mathcal{O}_Y(-H) \otimes \det(Q)^\nu) \otimes \mathcal{O}_{\mathbb{P}'}(\nu \cdot \gamma)$$

has a section. We put $\alpha = \nu \cdot \gamma$. Then we obtain a nontrivial map

$$\varphi : (\pi'_* \mathcal{O}_{\mathbb{P}'}(\alpha))^* = S^\alpha \left(\bigoplus^r \mathcal{E} \right) \to \mathcal{O}_Y(-H) \otimes \det(Q)^\nu.$$

By taking a birational modification $g : Y' \to Y$, we have

$$\mathcal{G} \otimes \mathcal{O}_{Y'}(F) = g^* \mathcal{O}_Y(-H) \otimes g^*(\det(Q)^\nu),$$

where F is an effective divisor on Y' and $\mathcal{G}$ is a quotient invertible sheaf of $g^*(S^\alpha(\bigoplus^r \mathcal{E}))$. Note that $\mathcal{G}$ is nef since $g^*(S^\alpha(\bigoplus^r \mathcal{E}))$ is nef (see Lemma 3.1.10). We put $n = \dim Y = \dim Y'$. Then we obtain

$$\begin{aligned}
(\det(Q)^{\nu})^n &= (g^* \det(Q)^{\nu})^n \\
&= (g^* \mathcal{O}_{Y'}(H) \otimes \mathcal{G} \otimes \mathcal{O}_{Y'}(F)) \cdot (g^* \det(Q)^{\nu})^{n-1} \\
&\geq g^* \mathcal{O}_{Y'}(H) \cdot (g^* \det(Q)^{\nu})^{n-1} \\
&= g^* \mathcal{O}_{Y'}(H) \cdot (g^* \mathcal{O}_{Y'}(H) \otimes \mathcal{G} \otimes \mathcal{O}_{Y'}(F)) \cdot (g^* \det(Q)^{\nu})^{n-2} \\
&\geq (g^* \mathcal{O}_{Y'}(H))^2 \cdot (g^* \det(Q)^{\nu})^{n-2} \\
&\geq \cdots \\
&\geq (g^* \mathcal{O}_{Y'}(H))^n \\
&= H^n > 0.
\end{aligned}$$

Here, we used the fact that $\det(Q) = \det f_* \mathcal{O}_X(\mu l K_{X/Y})$ is a nef invertible sheaf. Therefore, we see that $\det f_* \mathcal{O}_X(\mu l K_{X/Y})$ is a nef and big invertible sheaf on Y by

$$\left(\det f_* \mathcal{O}_X(\mu l K_{X/Y})\right)^{\dim Y} > 0.$$

Thus, we obtain that $\det f'_* \omega_{X'/Y'}^{\otimes k}$ is a nef and big invertible sheaf on Y' for some positive integer k. $\qquad\square$

Although we used some deep results, for example, the weak semistable reduction theorem by Abramovich–Karu (see Sect. 2.3.4 and [AbK]), the existence of relative canonical models for fiber spaces whose geometric generic fiber is of general type by Birkar–Cascini–Hacon–McKernan (see [BCHM]), in the proof of Theorem 4.3.1, we are released from the theory of variations of Hodge structure. We note that the proofs of Corollary 4.3.5 by Kawamata and Kollár depend on some deep results of the theory of variations of Hodge structure (see [Kaw6] and [Kol3]). We also note that the proof of Theorem 4.3.1 is independent of Viehweg's theory of weakly positive sheaves. We only need the nefness of some locally free sheaves in the proof of Theorem 4.3.1.

4.4 Elliptic Fibrations

Although the results in this section are more or less well known to the experts, we discuss elliptic fibrations for the reader's convenience. We will use Corollary 4.4.4 in the proof of Theorem 1.2.1 in Sect. 5.2. First, let us recall:

Theorem 4.4.1 ($\cdots$, Kawamata, Nakayama, $\cdots$) *Let* $f : V \to W$ *be a surjective morphism between smooth projective varieties whose general fibers are elliptic curves. Assume that there exists a simple normal crossing divisor* Σ *on* W *such that* f *is smooth over* $W_0 = W \setminus \Sigma$. *We further assume that all the local monodromies on* $R^1 f_{0*} \mathbb{C}_{V_0}$ *around* Σ *are unipotent, where* $f_0 = f|_{V_0} : V_0 = f^{-1}(W_0) \to W_0$. *Then we have*

$$(f_* \omega_{V/W})^{\otimes 12} \simeq J^* \mathcal{O}_{\mathbb{P}^1}(1),$$

*where $J : W \to \mathbb{P}^1$ is the natural extension of the period map $p : W_0 \to \mathbb{C} \simeq$
$\mathfrak{h}/SL(2, \mathbb{Z})$. Note that $\mathfrak{h} = \{z \in \mathbb{C} \,|\, \mathrm{Im}(z) > 0\}$.*

We do not prove this theorem here. Note that Theorem 4.4.1 is a special case of
[Kaw5, Theorem 20]. For a more detailed description of the period map $p : W_0 \to$
$\mathfrak{h}/SL(2, \mathbb{Z})$, see [N3, Corollary 3.2.1]. For a higher-dimensional generalization, see
[F3, Theorem 2.11], where we discuss period maps of polarized variations of Hodge
structure of weight one. Of course, Theorem 4.4.1 is also a special case of [F3,
Theorem 2.11].

4.4.2 Let $f : X \to Y$ be a projective surjective morphism between smooth projective varieties whose general fibers are elliptic curves. We can construct the following
commutative diagram:

$$
\begin{array}{ccc}
X & \xleftarrow{\ \rho\ } & X' \\
{\scriptstyle f}\downarrow & & \downarrow{\scriptstyle f'} \\
Y & \xleftarrow{\ \tau\ } & Y'
\end{array}
$$

such that:

(i) $\tau : Y' \to Y$ is a generically finite surjective morphism from a smooth projective
      variety Y',
(ii) X' is a smooth projective variety which is a resolution of the main component
      of $X \times_Y Y'$,
(iii) there exists a simple normal crossing divisor Σ on Y' such that f' is smooth over
      $Y'_0 = Y' \setminus \Sigma$, $f'_0 = f'|_{X'_0} : X'_0 = f'^{-1}(Y'_0) \to Y'_0$ has a section, $f'_0 : X'_0 \to Y'_0$
      is an elliptic curve with level 3-structure, and
(iv) all the local monodromies on $R^1 f'_{0*} \mathbb{C}_{X'_0}$ around Σ are unipotent.

For the details, see, for example, [KatM, Theorem 2.1.2, Theorem 3.7.1, and so on].
Let $M_1^{(3)}$ be the fine moduli scheme of elliptic curves with level 3-structure (see, for
example, [AbO, Theorem 13.1]). Note that $M_1^{(3)}$ is a finite cover of $\mathbb{C} = \mathfrak{h}/SL(2, \mathbb{Z})$.
Let $C \to M_1^{(3)}$ be the universal family. Then there exists a morphism $\alpha : Y'_0 \to M_1^{(3)}$
such that $X'_0 = C \times_{M_1^{(3)}} Y'_0$. By Theorem 4.4.1, we have the period map $p : Y'_0 \to$
$\mathbb{C} = \mathfrak{h}/SL(2, \mathbb{Z})$ and its extension $J : Y' \to \mathbb{P}^1$. We note the following commutative
diagram:

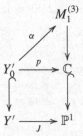

Therefore, we see

$$\mathrm{Var}(f) = \dim \overline{\alpha(Y_0')} = \dim J(Y').$$

By Theorem 4.4.1,

$$(f_*'\omega_{X'/Y'})^{\otimes 12} \simeq J^*\mathcal{O}_{\mathbb{P}^1}(1).$$

This implies that

$$\kappa(Y', \det f_*'\omega_{X'/Y'}) = \kappa(Y', f_*'\omega_{X'/Y'}) = \kappa(Y', J^*\mathcal{O}_{\mathbb{P}^1}(1)) = \mathrm{Var}(f).$$

By Theorem 2.3.46 (see [AbK, Theorem 0.3]) and Lemma 2.3.49 (see [AbK, Lemma 6.2]), we can take a generically finite morphism $\tau' : Y'' \to Y'$ from a smooth projective variety such that:

(v) $\mathrm{Supp}\,\tau'^*\Sigma$ is a simple normal crossing divisor on Y'',
(vi) there exists a projective birational morphism $X^\dagger \to X' \times_{Y'} Y''$ such that the induced morphism $f^\dagger : X^\dagger \to Y''$ is weakly semistable.

Let $X'' \to X^\dagger$ be a birational morphism from a smooth projective variety X'' such that $f'' : X'' \to Y''$ is the induced morphism. In this case, we see that

$$\tau'^* f_*'\omega_{X'/Y'} = f_*''\omega_{X''/Y''}.$$

This is because $f_*'\omega_{X'/Y'}$ is characterized as the canonical extension of a suitable Hodge bundle (see [Kaw3, Sect. 4. Semi-positivity (1)], [Kol2, Theorem 2.6], [N2, Theorem 1], and so on) and $\mathrm{Supp}\,\tau'^*\Sigma$ is a simple normal crossing divisor on Y'' (see [Kaw5, Proposition 1], and so on). For some more general results on the theory of variations of (mixed) Hodge structure, see [F4], [FF1], [FFS], and so on. Therefore, we have

$$\kappa(Y'', \det f_*''\omega_{X''/Y''}) = \kappa(Y'', f_*''\omega_{X''/Y''}) = \kappa(Y', f_*'\omega_{X'/Y'}) = \mathrm{Var}(f).$$

Moreover, by [F10, Theorem 1.6] (see also Remarks 4.3.12 and 4.3.13), we see that $f_*''\omega_{X''/Y''}^{\otimes m}$ is nef for every positive integer m.

Therefore, by the above observation, the Viehweg Conjecture Q holds true when general fibers of $f : X \to Y$ are elliptic curves by Remark 4.1.11, Theorem 3.3.12, and Remark 3.3.13.

Theorem 4.4.3 *Let $f : X \to Y$ be a surjective morphism between smooth projective varieties whose general fibers are elliptic curves. Assume that $\mathrm{Var}(f) = \dim Y$. Then $f_*\omega_{X/Y}^{\otimes k}$ is big for some positive integer k.*

By the above description of elliptic fibrations, we have:

Corollary 4.4.4 *Let $f : X \to Y$ be a surjective morphism between smooth projective varieties with connected fibers whose general fibers are elliptic curves. Then we have the following properties:*

(i) *We have*

$$\kappa(Y, (f_*\omega_{X/Y})^{**}) \geq \mathrm{Var}(f).$$

Note that $(f_*\omega_{X/Y})^{**}$ *is an invertible sheaf on* Y.

(ii) *Let* $\mathcal{L}$ *be an invertible sheaf on* Y. *If* $\kappa(Y, \mathcal{L}) \geq 0$, *then we have*

$$\kappa(X, \omega_{X/Y} \otimes f^*\mathcal{L}) \geq \kappa(X_{\overline{\eta}}) + \max\{\kappa(Y, \mathcal{L}), \mathrm{Var}(f)\}$$
$$= \max\{\kappa(Y, \mathcal{L}), \mathrm{Var}(f)\}.$$

(iii) *We have*

$$\kappa(X, \omega_{X/Y}) \geq \kappa(X_{\overline{\eta}}) + \mathrm{Var}(f)$$
$$= \mathrm{Var}(f).$$

(iv) *If* $\kappa(Y) \geq 0$, *then we have*

$$\kappa(X) \geq \kappa(X_{\overline{\eta}}) + \max\{\kappa(Y), \mathrm{Var}(f)\}$$
$$= \max\{\kappa(Y), \mathrm{Var}(f)\}.$$

Proof The statements (iii) and (iv) are important special cases of (ii). In the big commutative diagram constructed in Lemma 4.1.5, we can choose a weakly semistable morphism $g'' : V'' \to W''$ such that we can apply the result in 4.4.2 to $g'' : V'' \to W''$, that is, $\kappa(W'', g''_*\omega_{V''/W''}) = \mathrm{Var}(f)$. Note that $g''_*\omega_{V''/W''}$ is an invertible sheaf. Therefore, we obtain the desired inequality in (ii) by Lemma 4.1.7. We also obtain the desired inequality in (i) by Lemma 4.1.8. □

By combining Corollary 4.3.5 with Corollary 4.4.4, we have $C_{n,n-1}^+$ (cf. [Vi1] and [Vi2]).

Corollary 4.4.5 $(C_{n,n-1}^+)$ *Let* $f : X \to Y$ *be a surjective morphism between smooth projective varieties whose general fibers are irreducible curves. Assume that* $\kappa(Y) \geq 0$. *Then we have*

$$\kappa(X) \geq \kappa(X_y) + \max\{\mathrm{Var}(f), \kappa(Y)\},$$

where X_y *is a general fiber of* $f : X \to Y$.

4.5 Some Other Cases

In Sects. 4.2, 4.3, and 4.4, we proved the Iitaka Conjecture C for $f : X \to Y$ in the cases where Y is of general type, the geometric generic fiber $X_{\overline{\eta}}$ of f is of general type, and the general fibers of f are elliptic curves, respectively. In this section, we treat some other interesting cases of the Iitaka Conjecture C. In Sect. 4.5.1, we treat the case where the base space Y has maximal Albanese dimension. In [CaP],

Cao and Păun use some new analytic methods (see Sect. 3.4.1, [PăT], and [HPS]) to prove Theorem 4.5.1. In Sect. 4.5.2, we quickly see Kawamata's result in [Kaw6]. His result says that the Viehweg Conjecture Q follows from the good minimal model conjecture. Therefore, in some sense, the minimal model program supersedes the Iitaka program by Theorem 4.5.5 (see also Remark 4.5.6). Finally, in Sect. 4.5.3, we discuss some recent developments on the Iitaka conjecture in positive characteristic.

4.5.1 Fiber Spaces Whose Base Space Has Maximal Albanese Dimension

One of the most important recent results on the Iitaka conjecture is the following theorem by Cao–Păun.

Theorem 4.5.1 (Cao–Păun, see [CaP]) *Let* $f : X \to Y$ *be a surjective morphism from a smooth projective variety* X *to an Abelian variety* Y *with connected fibers. Then*

$$\kappa(X) \geq \kappa(X_y) + \kappa(Y)$$
$$= \kappa(X_y)$$

holds, where X_y *is a sufficiently general fiber of* $f : X \to Y$.

Theorem 3.4.3 plays a crucial role in the proof of Theorem 4.5.1. This means that we need some analytic methods for the proof of Theorem 4.5.1.

Let us recall the definition of maximal Albanese dimensional varieties.

Definition 4.5.2 (*Maximal Albanese dimensional varieties*) Let Y be a smooth projective variety and let $\alpha : Y \to \text{Alb}(Y)$ be the Albanese mapping. We say that Y has *maximal Albanese dimension* if $\alpha : Y \to \text{Alb}(Y)$ is generically finite over its image.

The following statement follows from Theorem 4.5.1 by standard arguments (see [HPS]).

Theorem 4.5.3 *Let* $f : X \to Y$ *be a surjective morphism between smooth projective varieties with connected fibers. Assume that* Y *has maximal Albanese dimension. Then*

$$\kappa(X) \geq \kappa(X_y) + \kappa(Y)$$

holds, where X_y *is a sufficiently general fiber of* f.

Let C be a smooth projective curve of genus ≥ 1. Then the Albanese mapping $\alpha : X \to \text{Alb}(C)$ is an embedding. Therefore, C has maximal Albanese dimension. Hence Theorem 4.5.3 recovers Kawamata's theorem.

Corollary 4.5.4 $(C_{n,1}$, [Kaw4, Theorem 2]) *Let* $f : X \to Y$ *be a surjective morphism from a smooth projective variety* X *to a smooth projective curve with connected fibers. Then*

$$\kappa(X) \geq \kappa(X_y) + \kappa(Y)$$

holds, where X_y *is a sufficiently general fiber of* $f : X \to Y$.

We strongly recommend the interested reader to see [HPS], which is very accessible to algebraic geometers. In [HPS], Hacon, Popa, and Schnell combine the theory of generic vanishing theorems, which is known to be very powerful for the study of maximal Albanese dimensional varieties, with some new analytic methods mainly due to [PăT] and [CaP] to make Theorem 4.5.1 more accessible to algebraic geometers.

4.5.2 Fiber Spaces Whose Geometric Generic Fiber Has a Good Minimal Model

Let us quickly see Kawamata's result in [Kaw6]. Here, we will use the standard notation of the minimal model program as in [F7] and [F12].

Theorem 4.5.5 ([Kaw6, Theorem 1.1]) *Let* $f : X \to Y$ *be a surjective morphism between smooth projective varieties with connected fibers and let* $\mathcal{L}$ *be an invertible sheaf on* Y. *Assume that the geometric generic fiber* $X_{\overline{\eta}}$ *has a good minimal model. Then the following assertions hold:*

(i) *There exists a positive integer* n *such that*

$$\kappa(Y, \widehat{\det}(f_* \omega_{X/Y}^{\otimes n})) \geq \text{Var}(f).$$

(ii) *If* $\kappa(Y, \mathcal{L}) \geq 0$, *then*

$$\kappa(X, \omega_{X/Y} \otimes f^* \mathcal{L}) \geq \kappa(X_{\overline{\eta}}) + \max\{\kappa(Y, \mathcal{L}), \text{Var}(f)\}.$$

Remark 4.5.6 By combining Theorem 4.5.5(i) with Theorem 3.3.12 and Remark 4.1.11, we see that Theorem 4.5.5 proves the Viehweg Conjecture Q under the assumption that the geometric generic fiber has a good minimal model.

As an obvious corollary of Theorem 4.5.5, we have:

Corollary 4.5.7 ([Kaw6, Corollary 1.2]) *Under the same assumptions and notation as in Theorem 4.5.5, we have:*

(i) $\kappa(X, \omega_{X/Y}) \geq \kappa(X_{\bar{\eta}}) + \mathrm{Var}(f)$, *and*

(ii) if $\kappa(Y) \geq 0$, *then*

$$\kappa(X) \geq \kappa(X_{\bar{\eta}}) + \max\{\kappa(Y), \mathrm{Var}(f)\}.$$

The proof of Theorem 4.5.5, which is Hodge theoretic, is beyond the scope of this book. For the details, see Kawamata's original paper: [Kaw6]. We note that [Kaw6, Theorem 5.5], which is nothing but [Kaw5, Theorem 3], has some troubles. We need [Kol3] to make the proof of [Kaw5, Theorem 3] rigorous. For the details, see [Kol3, I. Introduction] (see also [LD]).

Remark 4.5.8 Let V be a smooth projective variety. Then we have already known that V has a good minimal model if $\kappa(V) \geq \dim V - 3$ (see, for example, [Lai]). Therefore, the Viehweg Conjecture Q holds true under the assumption that $\kappa(X_{\bar{\eta}}) \geq \dim X_{\bar{\eta}} - 3$ by Theorem 4.5.5 and Remark 4.5.6.

We believe that the arguments in Sect. 4.3 are more accessible to algebraic geometers than [Kaw6] and [Kol3] when the geometric generic fiber is of general type.

4.5.3 Iitaka Conjecture in Positive Characteristic

Here, we will work over an algebraically closed field k of positive characteristic. We note that a *fibration* means a projective morphism $f : X \to Y$ between varieties such that the natural map $\mathcal{O}_Y \to f_*\mathcal{O}_X$ is an isomorphism.

Conjecture 4.5.9 (Conjecture $C_{n,m}$ in positive characteristic) *Let* $f : X \to Y$ *be a fibration between smooth projective varieties of dimension n, m respectively over an algebraically closed field k. Assume that the geometric generic fiber* $X_{\bar{\eta}}$ *is integral and smooth. Then*

$$\kappa(X) \geq \kappa(X_{\bar{\eta}}) + \kappa(Y).$$

There are many interesting results related to Conjecture 4.5.9, semipositivity, and weak positivity in positive characteristic (see, for example, [BCZ], [ChZ], [Ej], [EjZ], [Pat1], [Pat2], and [Z]). Here, we only make the following two statements.

Theorem 4.5.10 (Chen–Zhang, see [ChZ, Theorem 1.2]) *Conjecture* $C_{n,n-1}$ *holds for every n in any characteristic.*

We think that this result is not so surprising, since we can use the theory of moduli of curves in any characteristic.

Theorem 4.5.11 (Ejiri–Zhang, see [EjZ, Theorem 1.2]) *Conjecture* $C_{3,m}$ *holds for every m when* $\mathrm{char} k = p > 5$.

We note that Hacon and Xu started to study the minimal model program for 3-folds in characteristic $p > 5$ in [HX]. After [HX], the minimal model program for 3-folds in characteristic $p > 5$ has developed rapidly and drastically (see [Bi2], [BW], [Wa1], [Wa2], [HNT], and so on). We recommend the reader to see [HNT] for the current status of the minimal model program for 3-folds in characteristic $p > 5$. Since the proof of Theorem 4.5.11 uses the minimal model program for 3-folds, it needs the assumption that the characteristic is greater than 5.

In positive characteristic, many important conjectures are still widely open.

Chapter 5
$\overline{C}_{n,n-1}$ Revisited

This chapter is a revised version of the author's unpublished preprint [F1]. In this chapter, we give a detailed proof of the following theorem, which is Theorem 1.2.1 in this book.

Theorem ([Kaw2, Theorem 1], see Theorems 1.2.1 and 5.2.1) *Let $f : X \to Y$ be a dominant morphism of algebraic varieties defined over the complex number field $\mathbb{C}$. We assume that the general fiber $X_y = f^{-1}(y)$ is an irreducible curve. Then we have the following inequality for logarithmic Kodaira dimensions:*

$$\overline{\kappa}(X) \geq \overline{\kappa}(Y) + \overline{\kappa}(X_y).$$

This theorem says that the Iitaka Conjecture $\overline{C}_{n,n-1}$ holds true for every positive integer n. In Sect. 5.1, we see the background of this theorem. We explain the author's motivation for this chapter. Section 5.2 is the main part of this chapter. In Sect. 5.2, we give a detailed proof of Theorem 1.2.1 by using the weak semistable reduction theorem due to Abramovich–Karu (see [AbK]). Our proof is not the same as Kawamata's original one in [Kaw2]. In the proof of Theorem 1.2.1, we use some results established and explained in Chap. 4. In Sect. 5.3, we quickly explain some related results without proof for the interested readers. In Sect. 5.4, which is an appendix, we give a quick proof of [Kaw2, Lemma 4] based on the Kawamata–Viehweg vanishing theorem, although we do not need [Kaw2, Lemma 4] in this book. We hope that this chapter will make Theorem 1.2.1, that is, $\overline{C}_{n,n-1}$, more accessible.

5.1 Background of $\overline{C}_{n,n-1}$

Let us quote the introduction of [F1] for the reader's convenience.

© The Author(s), under exclusive license to Springer Nature Singapore Pte Ltd. 2020
O. Fujino, *Iitaka Conjecture*, SpringerBriefs in Mathematics,
https://doi.org/10.1007/978-981-15-3347-1_5

In spite of its importance, the proof of $\overline{C}_{n,n-1}$ is not so easy to access for the younger generation, including myself. After [Kaw2] was published, the birational geometry has drastically developed. When Kawamata wrote [Kaw2], the following techniques and results are not known or fully matured.

- Kawamata's covering trick,
- moduli theory of curves, especially, the notion of level structures and the existence of tautological families,
- various notions of singularities such as rational singularities, canonical singularities, and so on.

See [Kaw3, Sect. 2], [AbK, Sect. 5], [AbO, Part II], [vaGO], [Vi2], and [KolM]. In the mid 1990s, de Jong gave us fantastic results: [dJ1] and [dJ2]. The alteration paradigm generated the weak semistable reduction theorem [AbK]. This paper shows how to simplify the proof of the main theorem of [Kaw2] by using the weak semistable reduction. The proof may look much simpler than Kawamata's original proof (note that we have to read [Vi1] and [Vi2] to understand [Kaw2]). However, the alteration theorem grew out from the deep investigation of the moduli space of stable pointed curves (see [dJ1] and [dJ2]). So, don't misunderstand the real value of this paper. We note that we do not enforce Kawamata's arguments. We only recover his main result. Of course, this paper is not self-contained.

We think that it is much easier to give a rigorous proof of Theorem 1.2.1 without depending on Kawamata's paper [Kaw2] than to check all the details of [Kaw2] and correct some mistakes in [Kaw2]. We note that [Kaw2, Lemma 2] does not take Viehweg's correction [Vi2] into account.

5.1.1 (Background and motivation) In the proof of [Kaw2, Lemma 4], Kawamata considered the following commutative diagram:

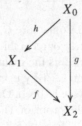

in order to prove $R^i f_* \mathcal{O}_{X_1}(-D_1) = 0$ for every $i > 0$. In the first half of the proof of [Kaw2, Lemma 4], he proved $R^i g_* \mathcal{O}_{X_0}(-D_0) = 0$ for every $i > 0$ by direct easy calculations. The author is not sure if Kawamata's argument in the proof of [Kaw2, Lemma 4] is sufficient for proving $R^i f_* \mathcal{O}_{X_1}(-D_1) = 0$ for every $i > 0$ from $R^i g_* \mathcal{O}_{X_0}(-D_0) = 0$ for every $i > 0$. We can check $R^i f_* \mathcal{O}_{X_1}(-D_1) = 0$ for $i > 0$ as follows.

Let us consider the usual spectral sequence:

$$E_2^{p,q} = R^p f_* R^q h_* \mathcal{O}_{X_0}(-D_0) \Rightarrow R^{p+q} g_* \mathcal{O}_{X_0}(-D_0).$$

Note that $h_* \mathcal{O}_{X_0}(-D_0) \simeq \mathcal{O}_{X_1}(-D_1)$ by the definitions of D_0 and D_1. Since

$$E_2^{1,0} \simeq R^1 f_* \mathcal{O}_{X_1}(-D_1) \hookrightarrow R^1 g_* \mathcal{O}_{X_0}(-D_0) = 0,$$

we obtain $R^1 f_* \mathcal{O}_{X_1}(-D_1) = 0$. By applying this argument to $h : X_0 \to X_1$, we can prove $R^1 h_* \mathcal{O}_{X_0}(-D_0) = 0$. This is a crucial step. This implies that $E_2^{p,1} = 0$ for every p. Thus we obtain the inclusion

$$E_2^{2,0} \simeq E_\infty^{2,0} \hookrightarrow R^2 g_* \mathcal{O}_{X_0}(-D_0) = 0.$$

Therefore, we have $E_2^{2,0} \simeq R^2 f_* \mathcal{O}_{X_1}(-D_1) = 0$. As above, we obtain $R^2 h_* \mathcal{O}_{X_0}(-D_0) = 0$ by applying this argument to $h : X_0 \to X_1$. This implies that $E_2^{p,1} = E_2^{p,2} = 0$ for every p. Then we get the inclusion

$$E_2^{3,0} \simeq E_\infty^{3,0} \hookrightarrow R^3 g_* \mathcal{O}_{X_0}(-D_0) = 0$$

and $E_2^{3,0} \simeq R^3 f_* \mathcal{O}_{X_1}(-D_1) = 0$. By repeating this process, we finally obtain that $R^i f_* \mathcal{O}_{X_1}(-D_1) = 0$ holds for every $i > 0$.

The author does not know whether the above understanding of [Kaw2, Lemma 4] is the same as what Kawamata wanted to say in the proof of [Kaw2, Lemma 4] or not. It seemed to the author that Kawamata only proves that the composition

$$\mathbf{R} f_*(\varphi_{01}) \circ \varphi_{12} : \mathcal{O}_{X_2}(-D_2) \to \mathbf{R} f_* \mathcal{O}_{X_1}(-D_1) \to \mathbf{R} f_* \mathbf{R} h_* \mathcal{O}_{X_0}(-D_0)$$

is a quasi-isomorphism in the derived category of coherent sheaves on X_2.

As we have already pointed out above, [Kaw2] does not take Viehweg's correction [Vi2] into account. Note that the statement of [Kaw2, Lemma 2], which claims that X is always Gorenstein if X is normal and there is a finite flat morphism $f : X \to Y$ onto a smooth variety Y, is wrong. This mistake comes from an error in [Vi1, Corollary 5.3]. Therefore, we have to correct the statement of [Kaw2, Lemma 2] and modify some related statements in [Kaw2] in order to complete the proof of Theorem 1.2.1 in [Kaw2].

The author gave up checking the technical details of [Kaw2] and correcting mistakes in [Kaw2], and decided to give a proof of Theorem 1.2.1 without depending on [Kaw2]. We will not use [Kaw2, Lemma 2] or [Kaw2, Lemma 4]. We will adopt a different approach to Theorem 1.2.1 in this book. The author believes that his decision is much more constructive. We also note that the reader does not have to refer to [Vi1] in order to understand the proof of Theorem 1.2.1 in this book. Therefore, the author thinks that the proof of Theorem 1.2.1 in this book is much more accessible than the original proof in [Kaw2].

5.2 $\overline{C}_{n,n-1}$

In this section, we give a proof of the following theorem (see Theorem 1.2.1), which is the main theorem of [Kaw2]. As was mentioned before, this section is a revised version of the author's unpublished short note [F1] written in 2003 in Princeton.

Theorem 5.2.1 ([Kaw2, Theorem 1]) *Let $f : V \to W$ be a dominant morphism of algebraic varieties defined over the complex number field $\mathbb{C}$. We assume that the general fiber $V_w = f^{-1}(w)$ is an irreducible curve. Then we have the following inequality for logarithmic Kodaira dimensions:*

$$\overline{\kappa}(V) \geq \overline{\kappa}(W) + \overline{\kappa}(V_w).$$

It is easy to see that this statement is equivalent to Theorem 5.2.2 by the basic properties of the logarithmic Kodaira dimension (see Lemma 2.3.36).

Theorem 5.2.2 ($\overline{C}_{n,n-1}$) *Let $f : V \to W$ be a surjective morphism between smooth projective varieties with connected fibers. Let C and D be simple normal crossing divisors on V and W respectively. We put $V_0 := V \setminus C$ and $W_0 := W \setminus D$. Assume that $f(V_0) \subset W_0$, equivalently, $\operatorname{Supp} f^* D \subset \operatorname{Supp} C$. Then the inequality*

$$\overline{\kappa}(V_0) \geq \overline{\kappa}(W_0) + \overline{\kappa}(F_0)$$

holds, where F_0 is a general fiber of $f_0 = f|_{V_0} : V_0 \to W_0$.

Precisely speaking, we will prove the following theorem in this section.

Theorem 5.2.3 ($\overline{C}'_{n,n-1}$) *Let $f : X \to Y$ be a surjective morphism between smooth projective varieties with connected fibers. Let C and D be simple normal crossing divisors on X and Y respectively. We put $X_0 := X \setminus C$ and $Y_0 := Y \setminus D$. Assume that $f(X_0) \subset Y_0$, equivalently, $\operatorname{Supp} f^* D \subset \operatorname{Supp} C$. Then the inequality*

$$\kappa(X, K_X + C - f^*(K_Y + D)) \geq \overline{\kappa}(F_0)$$

holds, where F_0 is a general fiber of $f_0 = f|_{X_0} : X_0 \to Y_0$.

We note:

Proposition 5.2.4 *Theorem 5.2.2 follows from Theorem 5.2.3.*

By this proposition, we see that Theorem 5.2.3 is sufficient for Theorem 5.2.1.

Proof of Proposition 5.2.4 For the proof of $\overline{C}_{n,n-1}$, we may assume that $\overline{\kappa}(W_0) = \kappa(W, K_W + D) \geq 0$ and $\overline{\kappa}(F_0) \geq 0$ in Theorem 5.2.2. Therefore, we have

$$\overline{\kappa}(V_0) = \kappa(V, K_V + C) \geq \kappa(V, K_V + C - f^*(K_W + D)) \geq \overline{\kappa}(F_0) \geq 0 \quad (5.2.1)$$

by $\kappa(W, K_W + D) \geq 0$ and Theorem 5.2.3. Since $\kappa(V, K_V + C - f^*(K_W + D)) \geq 0$ and $\kappa(W, K_W + D) \geq 0$, we have

$$\overline{\kappa}(V_0) = \kappa(V, K_V + C) \geq \kappa(W, K_W + D) = \overline{\kappa}(W_0) \geq 0. \qquad (5.2.2)$$

If $\overline{\kappa}(W_0) = 0$ or $\overline{\kappa}(F_0) = 0$, then the desired inequality

$$\overline{\kappa}(V_0) \geq \overline{\kappa}(W_0) + \overline{\kappa}(F_0)$$

holds true by (5.2.1) and (5.2.2), respectively. Therefore, from now on, we may further assume that $\overline{\kappa}(W_0) \geq 1$ and $\overline{\kappa}(F_0) = 1$. We take a sufficiently large and divisible positive integer m such that

$$H^0(V, \mathcal{O}_V(m(K_V + C) - f^*(m(K_W + D)))) \neq 0,$$

and that $\alpha = \Phi_{|m(K_V+C)|} : V \dashrightarrow \mathbb{P}^N$ and $\beta = \Phi_{|m(K_W+D)|} : Y \dashrightarrow \mathbb{P}^M$ give Iitaka fibrations of $K_V + C$ and $K_W + D$ respectively. Note that

$$0 \neq a \in H^0(V, \mathcal{O}_V(m(K_V + C) - f^*(m(K_W + D))))$$

gives an injection

$$\iota : H^0(W, \mathcal{O}_W(m(K_W + D))) \hookrightarrow H^0(V, \mathcal{O}_V(m(K_V + C))).$$

We consider the following commutative diagram:

$$
\begin{array}{ccccc}
V & \overset{\alpha}{\dashrightarrow} & V_m & \hookrightarrow & \mathbb{P}^N \\
{\scriptstyle f}\downarrow & & {\scriptstyle q}\downarrow & & \downarrow{\scriptstyle p} \\
W & \underset{\beta}{\dashrightarrow} & W_m & \hookrightarrow & \mathbb{P}^M,
\end{array}
$$

where V_m and W_m are the images of α and β respectively. We note that the projection $p : \mathbb{P}^N \dashrightarrow \mathbb{P}^M$ is induced by the inclusion ι. We assume that $\kappa(V, K_V + C) = \kappa(W, K_W + D)$. Then q is birational. By taking suitable birational modifications, we may assume that α and β are morphisms:

$$
\begin{array}{ccc}
V & \overset{\alpha}{\longrightarrow} & V_m \\
{\scriptstyle f}\downarrow & & \downarrow{\scriptstyle q} \\
W & \underset{\beta}{\longrightarrow} & W_m
\end{array}
$$

We take a sufficiently general point $P \in W_m$ and consider

$$
\begin{array}{ccc}
V \longleftarrow V' \longrightarrow P \\
f \downarrow \qquad f' \downarrow \qquad \| \\
W \longleftarrow W' \longrightarrow P
\end{array}
$$

where $V' = f^{-1}\beta^{-1}(P)$ and $W' = \beta^{-1}(P)$. We put $C' = C|_{V'}$ and $D' = D|_{W'}$. Then we have $\kappa(V', K_{V'} + C') = \kappa(W', K_{W'} + D') = 0$. By Theorem 5.2.3, we obtain

$$
0 = \kappa(V', K_{V'} + C') \geq \kappa(V', K_{V'} + C' - f'^*(K_{W'} + D'))
$$
$$
\geq \overline{\kappa}(F_0) = 1.
$$

This is a contradiction. Therefore, we obtain

$$
\overline{\kappa}(V_0) = \kappa(V, K_V + C) \geq \kappa(W, K_W + D) + 1 = \overline{\kappa}(W_0) + \overline{\kappa}(F_0).
$$

This is the desired inequality. □

Before we start the proof of Theorem 5.2.3, let us recall the following trivial lemma. We will frequently use it in the proof of Theorem 5.2.3 without mentioning it explicitly.

Lemma 5.2.5 *Let X be a normal projective variety. Let D_1 and D_2 be $\mathbb{Q}$-Cartier $\mathbb{Q}$-divisors on X. Assume that $D_1 \geq D_2$. Then we have $\kappa(X, D_1) \geq \kappa(X, D_2)$.*

Proof of Theorem 5.2.3 We divide the proof into several steps.

Step 1 We apply Theorem 2.3.44 (see [AbK, Theorem 2.1]) to $f : X \to Y$ and take some further projective birational modifications (see [AbK, Proposition 4.4 and Remark 4.5] and the proof of Theorem 2.3.46). Then we have the following commutative diagram:

$$
\begin{array}{ccc}
X \xleftarrow{p} X' \longleftarrow U_{X'} \\
f \downarrow \qquad f' \downarrow \qquad \downarrow \\
Y \xleftarrow{q} Y' \longleftarrow U_{Y'}
\end{array}
$$

where $p : X' \to X$ and $q : Y' \to Y$ are projective birational morphisms, X' has only quotient singularities, Y' is smooth, the inclusions on the right are toroidal embeddings without self-intersection with the following properties:

(i) $f' : (U_{X'} \subset X') \to (U_{Y'} \subset Y')$ is toroidal and equidimensional.
(ii) We put $C' := (p^*C)_{\text{red}}$ and $D' := (q^*D)_{\text{red}}$. Then $C' \subset X' \setminus U_{X'}$ and $D' \subset Y' \setminus U_{Y'}$.

For the details of (ii), see the construction in the proof of [AbK, Theorem 2.1] (see Remark 2.3.45). Note that [Kar1, Chap. 2, Sect. 9] is also helpful. Then we can write

$$K_{X'} + C' = p^*(K_X + C) + E$$

for some effective p-exceptional divisor E and

$$K_{Y'} + D' = q^*(K_Y + D) + F$$

for some effective q-exceptional divisor F. Therefore, we obtain

$$\overline{\kappa}(X_0) = \kappa(X, K_X + C) = \kappa(X', K_{X'} + C')$$

and

$$\overline{\kappa}(Y_0) = \kappa(Y, K_Y + D) = \kappa(Y', K_{Y'} + D').$$

Since

$$K_{X'} + C' - f'^*(K_{Y'} + D') = p^*(K_X + C - f^*(K_Y + D)) + E - f'^*F,$$

we have

$$\kappa(X, K_X + C - f^*(K_Y + D)) \geq \kappa(X', K_{X'} + C' - f'^*(K_{Y'} + D')),$$

Thus, we may replace $f : X \to Y$ with $f' : X' \to Y'$. From now on, we omit the superscript $'$ for simplicity of notation. So, we may assume that $f : X \to Y$ is toroidal with the above extra assumptions.

Step 2 By taking a suitable Kawamata cover $q : Y' \to Y$ (see Lemma 2.3.38), we obtain the following commutative diagram:

$$
\begin{array}{ccc}
X & \xleftarrow{\ p\ } & X' \\
{\scriptstyle f}\downarrow & & \downarrow{\scriptstyle f'} \\
Y & \xleftarrow{\ q\ } & Y'
\end{array}
$$

such that $f' : X' \to Y'$ is weakly semistable, where X' is the normalization of $X \times_Y Y'$ (see [AbK, Sect. 5]). Note that X' is Gorenstein by Lemma 2.3.43 (see [AbK, Lemma 6.1]). We put $G := X \setminus U_X$ and $H := Y \setminus U_Y$. Then we have

$$K_X + C - f^*(K_Y + D) \geq K_X + C_{\text{hor}} + G_{\text{ver}} - f^*(K_Y + H).$$

Therefore, we can check that

$$p^*(K_X + C - f^*(K_Y + D)) \geq K_{X'/Y'} + (p^*C)_{\text{hor}}.$$

We note that $(p^*C)_{hor} = p^*(C_{hor})$. So, it is sufficient to prove that $\kappa(X', K_{X'/Y'} + (p^*C)_{hor}) \geq \overline{\kappa}(F_0)$.

Step 3 Let F be a general fiber of $f : X \to Y$. We put $g := g(F)$: the genus of F.

Case 1 $(g \geq 2)$ In this case,

$$\kappa(X', K_{X'/Y'} + (p^*C)_{hor}) \geq \kappa(X', K_{X'/Y'}) \geq 1 = \overline{\kappa}(F_0)$$

by Corollary 4.3.5 (iii). Precisely speaking,

$$\kappa(X', K_{X'/Y'}) = \kappa(\widetilde{X}, K_{\widetilde{X}/Y'}),$$

where $\widetilde{X} \to X'$ is a resolution of singularities, since X' has only rational Gorenstein singularities, and $\kappa(\widetilde{X}, K_{\widetilde{X}/Y'}) \geq 1$ by Corollary 4.3.5 (iii).

Case 2 $(g = 1)$ By the description in 4.4.2 and Corollary 4.4.4, we have

$$\kappa(X', K_{X'/Y'}) \geq \text{Var}(f') = \text{Var}(f) \geq 0$$

since X' has only rational Gorenstein singularities as above. So, if C is vertical or $\text{Var}(f) \geq 1$, then we obtain

$$\kappa(X', K_{X'/Y'} + (p^*C)_{hor}) \geq \overline{\kappa}(F_0).$$

Therefore, we may assume that $\text{Var}(f) = 0$ and C is not vertical. Since $\text{Var}(f) = 0$, there is a generically finite surjective morphism $\tau : Y'' \to Y'$ from a normal projective variety Y'' such that $X'' = X' \times_{Y'} Y''$ is birationally equivalent to $Y'' \times E$, where E is an elliptic curve.

Lemma 5.2.6 *Let $\overline{\tau} : \overline{Y} \to Y'$ be a birational morphism from a smooth projective variety $\overline{Y}$ such that $\overline{\tau}^{-1}(Y' \setminus U_{Y'})$, where $(U_{Y'} \subset Y')$ is the toroidal structure of Y', is a simple normal crossing divisor on $\overline{Y}$. We have the following commutative diagram:*

$$
\begin{array}{ccc}
X' & \xleftarrow{\ \overline{\pi}\ } & \overline{X} \\
{\scriptstyle f'}\big\downarrow & & \big\downarrow{\scriptstyle \overline{f}} \\
Y' & \xleftarrow[\ \overline{\tau}\ ]{} & \overline{Y}
\end{array}
$$

where $\overline{X} = X' \times_{Y'} \overline{Y}$. Then $\overline{f} : \overline{X} \to \overline{Y}$ is weakly semistable and

$$\kappa(X', K_{X'/Y'} + (p^*C)_{hor}) \geq \kappa(\overline{X}, K_{\overline{X}/\overline{Y}} + (\overline{\pi}^* p^*C)_{hor}).$$

Proof of Lemma 5.2.6 Note that $\overline{f} : \overline{X} \to \overline{Y}$ is weakly semistable by Lemma 2.3.49 (see also [AbK, Lemma 6.2]). We also note that

$$K_{\overline{Y}} = \overline{\tau}^* K_{Y'} + E$$

and

$$K_{\overline{X}} = \overline{\pi}^* K_{X'} + F,$$

where E is an effective $\overline{\tau}$-exceptional divisor on $\overline{Y}$ and F is an effective $\overline{\pi}$-exceptional divisor on $\overline{X}$. Therefore, we obtain

$$K_{\overline{X}/\overline{Y}} + (\overline{\pi}^* p^* C)_{\mathrm{hor}} \le \overline{\pi}^* K_{X'/Y'} + F + \overline{\pi}^* (p^* C)_{\mathrm{hor}}.$$

This implies that the desired inequality

$$\kappa(X', K_{X'/Y'} + (p^* C)_{\mathrm{hor}}) \ge \kappa(\overline{X}, K_{\overline{X}/\overline{Y}} + (\overline{\pi}^* p^* C)_{\mathrm{hor}})$$

holds. □

By modifying Y' birationally, we may assume that there exists a simple normal crossing divisor Σ on Y' such that $\tau : Y'' \to Y'$ is finite étale over $Y' \setminus \Sigma$ (see Lemma 5.2.6). We may assume that $\tau : Y'' \to Y'$ is finite by using the Stein factorization. By Lemma 2.3.39, we may further assume that Y'' is a smooth projective variety. Therefore, we obtain the following commutative diagram:

$$
\begin{array}{ccc}
X' & \xleftarrow{\ \pi\ } & X'' \\
{\scriptstyle f'}\downarrow & & \downarrow{\scriptstyle f''} \\
Y' & \xleftarrow{\ \tau\ } & Y''
\end{array}
$$

where $\tau : Y'' \to Y'$ is a finite cover from a smooth projective variety Y'', $f'' : X'' := X' \times_{Y'} Y'' \to Y''$ is weakly semistable, and f'' is birationally equivalent to $Y'' \times E \to Y''$. Since

$$\pi^*(K_{X'/Y'} + (p^* C)_{\mathrm{hor}}) = K_{X''/Y''} + \pi^*((p^* C)_{\mathrm{hor}}),$$

it is sufficient to prove $\kappa(X'', K_{X''/Y''} + \pi^*((p^* C)_{\mathrm{hor}})) \ge 1$. Let $\alpha : \widetilde{X} \to Y'' \times E$ and $\beta : \widetilde{X} \to X''$ be a common resolution:

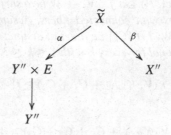

Since X'' has only rational Gorenstein singularities, X'' has at worst canonical Goren-stein singularities (see Lemma 2.3.13). Thus, we obtain

$$\kappa(X'', K_{X''/Y''} + \pi^*((p^*C)_{\text{hor}})) = \kappa(\widetilde{X}, K_{\widetilde{X}/Y''} + \beta^*\pi^*((p^*C)_{\text{hor}})).$$

On the other hand,

$$K_{\widetilde{X}/Y''} = K_{\widetilde{X}/Y''\times E} + K_{Y''\times E/Y''} =: A$$

is an effective α-exceptional divisor such that $\text{Supp}\,A = \text{Exc}(\alpha)$. Let B be an irre-ducible component of $\beta^*\pi^*((p^*C)_{\text{hor}})$ such that B is dominant onto Y''. Then

$$m(A + \beta^*\pi^*((p^*C)_{\text{hor}})) \geq \alpha^*\alpha_*B,$$

for a sufficiently large integer m. Therefore, it is sufficient to prove $\kappa(Y'' \times E, \alpha_*B) \geq 1$. This holds true by Corollary 2.3.54. Thus, we finish the proof when $g = 1$.

Case 3 $(g = 0)$ In this case, the general fiber of $f : X \to Y$ is $\mathbb{P}^1$. We take a general normal irreducible divisor H on X' which is horizontal with respect to $f' : X' \to Y'$. We consider the base change of $f' : X' \to Y'$ by $f'|_H : H \to Y'$. Then $X' \times_{Y'} H$ is birationally equivalent to $H \times \mathbb{P}^1$ by construction. As in the above case, after modifying Y' birationally, we can take a finite cover $\tau : Y'' \to Y'$ and obtain the following commutative diagram:

$$
\begin{array}{ccc}
X' & \xleftarrow{\;\pi\;} & X'' \\
{\scriptstyle f'}\downarrow & & \downarrow{\scriptstyle f''} \\
Y' & \xleftarrow{\;\tau\;} & Y''
\end{array}
$$

such that $f'' : X'' := X' \times_{Y'} Y'' \to Y''$ is weakly semistable and that $f'' : X'' \to Y''$ is birationally equivalent to $Y'' \times \mathbb{P}^1$. If necessary, we repeat this argument. Then we may further assume that all the horizontal components of $\pi^*((p^*C)_{\text{hor}})$ are mapped onto Y'' birationally. As before, it is sufficient to prove

$$\kappa(X'', K_{X''/Y''} + \pi^*((p^*C)_{\text{hor}})) \geq \overline{\kappa}(F_0).$$

Lemma 5.2.7 (cf. [F3, Sect. 7]) *Let $f : V \to W$ be a surjective morphism between smooth projective varieties with connected fibers. Assume that f is birationally equivalent to $W \times \mathbb{P}^1 \to W$. Let $\{C_k\}$ be a set of distinct irreducible divisors such that $f : C_k \to W$ is birational for every k with $1 \leq k \leq 3$. Then*

$$\kappa(V, K_{V/W} + C_1 + C_2) \geq 0$$

and

$$\kappa(V, K_{V/W} + C_1 + C_2 + C_3) \geq 1.$$

Proof of Lemma 5.2.7 It is sufficient to prove that

$$\dim_{\mathbb{C}} H^0(V, \mathcal{O}_V(K_{V/W} + C_1 + C_2)) \geq 1$$

and

$$\dim_{\mathbb{C}} H^0(V, \mathcal{O}_V(K_{V/W} + C_1 + C_2 + C_3)) \geq 2.$$

Let us consider the following commutative diagram:

$$
\begin{array}{ccc}
V & \xleftarrow{\ \pi\ } & V' \\
{\scriptstyle f}\downarrow & & \downarrow{\scriptstyle f'} \\
W & \xleftarrow{\ \tau\ } & W'
\end{array}
$$

where π and τ are projective birational morphisms, and V' and W' are smooth projective varieties. Let C_k' be the strict transform of C_k on V' for every k. Then we have

$$H^0(V', \mathcal{O}_{V'}(K_{V'/W'} + C_1' + C_2')) \subset H^0(V, \mathcal{O}_V(K_{V/W} + C_1 + C_2))$$

and

$$H^0(V', \mathcal{O}_{V'}(K_{V'/W'} + C_1' + C_2' + C_3')) \subset H^0(V, \mathcal{O}_V(K_{V/W} + C_1 + C_2 + C_3)).$$

Therefore, by modifying V and W birationally and replacing C_k with its strict transform, we may assume that there exists a simple normal crossing divisor Σ on W such that

$$\varphi_{ij} : V_0 := f^{-1}(W_0) \simeq W_0 \times \mathbb{P}^1$$

with $\varphi_{ij}(C_i|_{V_0}) = W_0 \times \{0\}$ and $\varphi_{ij}(C_j|_{V_0}) = W_0 \times \{\infty\}$ for $i \neq j$, where $W_0 := W \setminus \Sigma$. We may further assume that there exists $\psi_{ij} : V \to \mathbb{P}^1$ such that $\psi_{ij}|_{V_0} = p_2 \circ \varphi_{ij}$, where p_2 is the second projection $W_0 \times \mathbb{P}^1 \to \mathbb{P}^1$. We may also assume that $\bigcup_k C_k \cup \mathrm{Supp}\, f^*\Sigma$ is a simple normal crossing divisor on V. Then we obtain

$$
\begin{aligned}
\wedge\, \psi_{ij}{}^*&\left(\frac{dz}{z}\right) \\
&\in \mathrm{Hom}_{\mathcal{O}_V}(f^*\mathcal{O}_W(K_W + \Sigma), \mathcal{O}_V(K_V + C_i + C_j + (f^*\Sigma)_{\mathrm{red}})) \\
&\simeq H^0(V, \mathcal{O}_V(K_{V/W} + C_i + C_j + (f^*\Sigma)_{\mathrm{red}} - f^*\Sigma)) \\
&\subset H^0(V, \mathcal{O}_V(K_{V/W} + C_i + C_j))
\end{aligned}
$$

for $i \neq j$, where z denotes a suitable inhomogeneous coordinate of $\mathbb{P}^1$. Therefore, we have

$$\dim_{\mathbb{C}} H^0(V, \mathcal{O}_V(K_{V/W} + C_1 + C_2)) \geq 1$$

and
$$\dim_{\mathbb{C}} H^0(V, \mathcal{O}_V(K_{V/W} + C_1 + C_2 + C_3)) \geq 2.$$

Thus, we obtain the required result.   □

Apply Lemma 5.2.7 to $\widetilde{X} \to Y''$, where $\beta : \widetilde{X} \to X''$ is a resolution of X''. Then we obtain
$$\kappa(\widetilde{X}, K_{\widetilde{X}/Y''} + \beta^*\pi^*((p^*C)_{\text{hor}})) \geq \overline{\kappa}(F_0).$$

Thus, we complete the proof of Theorem 5.2.3.   □

5.3 Some Related Results

In this section, we review some new results on the Iitaka Conjecture $\overline{C}$ without proof.

5.3.1 Conjecture $\overline{C}_{n,m}$ for Affine Varieties, and so on

We will quickly see some results in [F13] and [F18].

Theorem 5.3.1 ([F13, Theorem 1.3] and [F18, Theorem 2.1]) *Let $f : X \to Y$ be a surjective morphism between smooth projective varieties with connected fibers. Let D_X and D_Y be simple normal crossing divisors on X and Y, respectively. Assume that $\text{Supp} f^* D_Y \subset \text{Supp} D_X$. Then we have*
$$\kappa_\sigma(X, K_X + D_X) \geq \kappa_\sigma(F, K_F + D_X|_F) + \kappa(Y, K_Y + D_Y)$$
and
$$\kappa_\sigma(X, K_X + D_X) \geq \kappa(F, K_F + D_X|_F) + \kappa_\sigma(Y, K_Y + D_Y),$$
where F is a sufficiently general fiber of $f : X \to Y$. We note that κ_σ denotes Nakayama's numerical dimension in Definition 2.3.58.

Theorem 5.3.1 says that Conjecture 1.2.2 follows from the generalized abundance conjecture (see Conjecture 2.3.59). We use Nakayama's theory of ω-sheaves and $\widehat{\omega}$-sheaves (see Sect. 3.4.2) for the proof of Theorem 5.3.1.

As a Corollary of Theorem 5.3.1, we can prove Conjecture 1.2.2 for affine varieties by using the minimal model program.

Corollary 5.3.2 ([F13, Corollary 1.6]) *Let $g : V \to W$ be a dominant morphism from an affine variety V to a variety W whose general fibers are irreducible. Then*
$$\overline{\kappa}(V) \geq \overline{\kappa}(W) + \overline{\kappa}(V_w)$$

holds, where V_w is a sufficiently general fiber of $g : V \to W$.

For the details of Theorem 5.3.1 and Corollary 5.3.2, see [F13], [F18], and [F19].

5.3.2 Subadditivity Due to Kovács–Patakfalvi and Hashizume

Here, let us see some results in [KovP] and [Has2].

Theorem 5.3.3 ([KovP, Theorem 9.6]) *Let $f : X \to Y$ be a surjective morphism between smooth projective varieties with connected fibers. Let D_X and D_Y be effective $\mathbb{Q}$-divisors on X and Y, respectively. Assume that (X, D_X) and (Y, D_Y) are lc, $\mathrm{Supp} D_X$ and $\mathrm{Supp} D_Y$ are simple normal crossing divisors, and $K_X + D_X$ is big over Y. We further assume that either:*

- *both D_X and D_Y are reduced and $\mathrm{Supp} f^* D_Y \subset \mathrm{Supp} D_X$, or*
- *$D_X \geq f^* D_Y$.*

Then
$$\kappa(X, K_X + D_X) \geq \kappa(Y, K_Y + D_Y) + \kappa(F, K_F + D_X|_F)$$
$$= \kappa(Y, K_Y + D_Y) + \dim X - \dim Y$$

holds, where F is a sufficiently general fiber of $f : X \to Y$.

Corollary 5.3.4 ([KovP, Corollary 9.8]) *Let $g : V \to W$ be a dominant morphism between varieties whose general fibers are irreducible. Assume that $\overline{\kappa}(V_w) = \dim V_w$, where V_w is a sufficiently general fiber of $g : V \to W$. Then*

$$\overline{\kappa}(V) \geq \overline{\kappa}(W) + \overline{\kappa}(V_w)$$
$$= \overline{\kappa}(W) + \dim V - \dim W$$

holds, that is, Conjecture 1.2.2 is true under the assumption that $\overline{\kappa}(V_w) = \dim V_w$.

For the details, we recommend the reader to see [KovP].

By combining some results in [KovP] with the recent developments of the minimal model program, Hashizume obtained the following result.

Theorem 5.3.5 ([Has2, Theorem 1.2]) *Let $f : X \to Y$ be a surjective morphism between smooth projective varieties with connected fibers. Let D_X and D_Y be simple normal crossing divisors on X and Y, respectively. Assume that $\mathrm{Supp} f^* D_Y \subset \mathrm{Supp} D_X$. Let F be a sufficiently general fiber of f. We further assume that the equality $\kappa_\sigma(F, K_F + D_F) = \kappa(F, K_F + D_F)$ holds, where $K_F + D_F = (K_X + D_X)|_F$. Then the inequality*

$$\kappa(X, K_X + D_X) \geq \kappa(Y, K_Y + D_Y) + \kappa(F, K_F + D_F)$$

holds.

By using this theorem, Hashizume proved:

Corollary 5.3.6 ([Has2, Corollary 1.4]) *Let $g : V \to W$ be a dominant morphism between varieties whose general fibers are irreducible. Assume that* $\dim V - \dim W \leq$ *3 or* $\dim V \leq 5$. *Then*

$$\overline{\kappa}(V) \geq \overline{\kappa}(W) + \overline{\kappa}(V_w)$$

holds, where V_w is a sufficiently general fiber of $g : V \to W$. This means that Conjecture 1.2.2 is true under the assumption that $\dim V - \dim W \leq 3$ *or* $\dim V \leq 5$.

For the details of the above statements and some more general statements, see [Has2].

5.4 Appendix: A Vanishing Lemma

In this appendix, we see that we can quickly recover [Kaw2, Lemma 4] by the Kawamata–Viehweg vanishing theorem.

Lemma 5.4.1 ([Kaw2, Lemma 4]) *Let $f : X_1 \to X_2$ be a birational morphism of smooth complete varieties and let D_1 and D_2 be simple normal crossing divisors on X_1 and X_2, respectively. We assume that $D_1 = \operatorname{Supp} f^* D_2$. Then we have*

$$Rf_* \mathcal{O}_{X_1}(-D_1) \simeq \mathcal{O}_{X_2}(-D_2).$$

Proof Note that $D_1 = \lceil \varepsilon f^* D_2 \rceil$ for $0 < \varepsilon \ll 1$. Since f is birational, $\varepsilon f^* D_2$ is f-nef and f-big. Therefore, by the relative Kawamata–Viehweg vanishing theorem, we have $R^i f_* \mathcal{O}_{X_1}(K_{X_1} + D_1) = 0$ for every $i > 0$ (see, for example, [F12, Theorem 3.2.1]). We write

$$K_{X_1} = f^*(K_{X_2} + D_2) + \sum_E a(E, X_2, D_2)E.$$

Since $K_{X_2} + D_2$ is Cartier, $a(E, X_2, D_2) \in \mathbb{Z}$ for every E. Since X_2 is smooth and D_2 is a simple normal crossing divisor on X_2, it is well known that $a(E, X_2, D_2) \geq -1$ for every E, that is, (X_2, D_2) is lc (see, for example, [F12, Lemma 2.3.9]). We can easily see that $f(E) \subset \operatorname{Supp} D_2$ if $a(E, X_2, D_2) = -1$. Therefore, we obtain

$$K_{X_1} + D_1 = f^*(K_{X_2} + D_2) + F$$

for some effective f-exceptional Cartier divisor F on X_1. Thus, we obtain $f_* \mathcal{O}_{X_1}(K_{X_1} + D_1) \simeq \mathcal{O}_{X_2}(K_{X_2} + D_2)$. This means that

$$Rf_* \mathcal{O}_{X_1}(K_{X_1} + D_1) \simeq \mathcal{O}_{X_2}(K_{X_2} + D_2).$$

By Grothendieck duality, we have

$$
\begin{aligned}
Rf_*\mathcal{O}_{X_1}(-D_1) &\simeq R\mathcal{H}om(Rf_*\mathcal{O}_{X_1}(K_{X_1}+D_1), \mathcal{O}_{X_2}(K_{X_2})) \\
&\simeq \mathcal{H}om(\mathcal{O}_{X_2}(K_{X_2}+D_2), \mathcal{O}_{X_2}(K_{X_2})) \\
&\simeq \mathcal{O}_{X_2}(-D_2).
\end{aligned}
$$

This is the desired quasi-isomorphism. □

Note that the Kawamata–Viehweg vanishing theorem was not known when Kawamata wrote [Kaw2]. The reader can find various formulations and some generalizations of the Kawamata–Viehweg vanishing theorem in [F12]. We also note that we did not use [Kaw2, Lemma 4] in the proof of Theorem 5.2.3 in Sect. 5.2.

Chapter 6
Appendices

Sections 6.1 and 6.2 are appendices. In Sect. 6.1, we give a simple proof of $C_{2,1}$ based on the Enriques–Kodaira classification of smooth projective surfaces for the reader's convenience. In Sect. 6.2, we show that Iitaka's Conjecture C_n follows from weaker conjectures: Conjectures $(\geq)_n$ and $(>)_n$. Although we do not use the results in Sect. 6.2 explicitly in this book, the arguments are useful for the study of the Iitaka conjecture.

6.1 $C_{2,1}$

In this section, we show that $C_{2,1}$ holds true with the aid of the Enriques–Kodaira classification of smooth projective surfaces (see, for example, [BHPV, Chap. VI]).

Theorem 6.1.1 ($C_{2,1}$) *Let $f : X \to Y$ be a surjective morphism from a smooth projective surface X onto a smooth projective curve Y with connected fibers. Then the inequality*

$$\kappa(X) \geq \kappa(F) + \kappa(Y)$$

holds, where F is a general fiber of $f : X \to Y$.

Proof Let us check the inequality

$$\kappa(X) \geq \kappa(F) + \kappa(Y). \tag{6.1.1}$$

Case 1 ($\kappa(X) = -\infty$) We assume that $\kappa(X) = -\infty$.

In this case, it is sufficient to prove that $\kappa(F) = -\infty$ holds under the assumption that $\kappa(Y) \geq 0$. Let us consider the following commutative diagram of Albanese mappings:

© The Author(s), under exclusive license to Springer Nature Singapore Pte Ltd. 2020
O. Fujino, *Iitaka Conjecture*, SpringerBriefs in Mathematics,
https://doi.org/10.1007/978-981-15-3347-1_6

$$X \xrightarrow{\alpha_X} \mathrm{Alb}(X)$$
$$\downarrow f \qquad\qquad \downarrow$$
$$Y \hookrightarrow_{\alpha_Y} \mathrm{Alb}(Y)$$

We note that $\alpha_Y : Y \to \mathrm{Alb}(Y)$ is an embedding since Y is a curve. For the details of Albanese mappings, see [U1, Sect. 9]. Then we see that X is not rational and that the general fiber of α_X is $\mathbb{P}^1$ with the aid of the Enriques–Kodaira classification. This implies that $\kappa(F) = -\infty$ holds.

Case 2 ($\kappa(X) = 0$) We assume that $\kappa(X) = 0$.

In this case, it is sufficient to prove that $\kappa(F) = \kappa(Y) = 0$ under the assumption that $\kappa(F) \geq 0$ and $\kappa(Y) \geq 0$. We note that

$$1 \leq g(Y) = \dim_{\mathbb{C}} H^1(Y, \mathcal{O}_Y) \leq \dim_{\mathbb{C}} H^1(X, \mathcal{O}_X) = q(X)$$

holds. Therefore, X is birationally equivalent to an Abelian surface or a hyperelliptic surface by the Enriques–Kodaira classification. By contracting (-1)-curves on X, we may assume that there are no (-1)-curves on X. Therefore, we have $K_X \sim_{\mathbb{Q}} 0$. In particular, X is an Abelian surface or a hyperelliptic surface. Let us consider the following commutative diagram of Albanese mappings again:

$$X \xrightarrow{\alpha_X} \mathrm{Alb}(X)$$
$$\downarrow f \qquad\qquad \downarrow$$
$$Y \hookrightarrow_{\alpha_Y} \mathrm{Alb}(Y)$$

If X is an Abelian surface, then $\alpha_X : X \to \mathrm{Alb}(X)$ is an isomorphism. This implies that Y is an elliptic curve and $\alpha_Y : Y \to \mathrm{Alb}(Y)$ is also an isomorphism. Thus we see that F is an elliptic curve. Therefore, we get $\kappa(F) = \kappa(Y) = 0$. If X is a hyperelliptic surface, then α_X is surjective and the general fiber of α_X is an elliptic curve. Thus we see that $f : X \to Y$ coincides with $\alpha_X : X \to \mathrm{Alb}(X)$. In particular, Y is an elliptic curve. Therefore, we have $\kappa(F) = \kappa(Y) = 0$.

Case 3 ($\kappa(X) = 1$) We assume that $\kappa(X) = 1$.

In this case, there exists a morphism $\varphi : X \to C$ onto a smooth projective curve C with connected fibers such that the general fiber E of φ is an elliptic curve. If $f(E) = Y$, then we have $\kappa(Y) \leq 0$. If $f(E)$ is a point, then we see that F is an elliptic curve. This means that $\kappa(F) = 0$. Therefore, the inequality (6.1.1) holds true in this case.

Case 4 ($\kappa(X) = 2$) We assume that $\kappa(X) = 2$.

In this case, the inequality (6.1.1) is obvious.

We obtained that $C_{2,1}$ holds true. $\square$

In this book, we saw that $Q_{n,n-1}$ holds true (for the details, see Sects. 4.3 and 4.4). Therefore, we have $C_{2,1}^+$.

Theorem 6.1.2 ($C_{2,1}^+$, see Corollaries 4.3.5 and 4.4.4) *Let* $f : X \to Y$ *be a surjective morphism from a smooth projective surface* X *onto a smooth projective curve* Y *with connected fibers. Assume that* $\kappa(Y) \geq 0$. *Then the inequality*

$$\kappa(X) \geq \kappa(F) + \max\{\mathrm{Var}(f), \kappa(Y)\}$$

holds, where F *is a general fiber of* $f : X \to Y$ *and* $\mathrm{Var}(f)$ *denotes Viehweg's variation of* $f : X \to Y$.

Proof If $\kappa(F) = -\infty$, then the inequality is obvious. If $\kappa(F) = 0$, then F is an elliptic curve. In this case, the inequality follows from Corollary 4.4.4. If $\kappa(F) = 1$, then F is a smooth projective curve of general type. In this case, the geometric generic fiber $X_{\overline{\eta}}$ of $f : X \to Y$ is of general type. Thus the inequality follows from Corollary 4.3.5. $\qquad\square$

6.2 Conjectures $(\geq)_n$ and $(>)_n$

In this section, we closely follow [I4, Lecture 4]. The main result of this section is Theorem 6.2.5. Although we do not use it explicitly in this book, the arguments in the proof of Theorem 6.2.5 are very useful. So we give a detailed proof of Theorem 6.2.5.

Let us recall the Iitaka Conjecture 1.2.4.

Conjecture 6.2.1 (Conjecture C_n) *Let* $f : X \to Y$ *be a surjective morphism between smooth projective varieties with* $\dim X = n$. *Let* X_y *be a sufficiently general fiber of* $f : X \to Y$. *Then*

$$\kappa(X) \geq \kappa(X_y) + \kappa(Y)$$

holds.

The following two conjectures are special cases of Conjecture 6.2.1, which may look obvious.

Conjecture 6.2.2 (Conjecture $(\geq)_n$) *Let* $f : X \to Y$ *be a surjective morphism between smooth projective varieties with* $\dim X = n$. *Let* X_y *be a sufficiently general fiber of* $f : X \to Y$. *If* $\kappa(X_y) \geq 0$ *and* $\kappa(Y) \geq 0$, *then*

$$\kappa(X) \geq 0$$

holds.

Conjecture 6.2.3 (Conjecture $(>)_n$) *Let $f : X \to Y$ be a surjective morphism between smooth projective varieties with $\dim X = n$. Let X_y be a sufficiently general fiber of $f : X \to Y$. If $\kappa(X_y) + \kappa(Y) > 0$, then*

$$\kappa(X) > 0$$

holds.

Remark 6.2.4 Conjectures $(\geq)_k$ and $(>)_k$ with $k \leq n$ easily follow from Conjectures $(\geq)_n$ and $(>)_n$, respectively. This is because $\kappa(X \times A) = \kappa(X)$ holds for every Abelian variety A.

In this section, we prove:

Theorem 6.2.5 *If Conjectures $(\geq)_n$ and $(>)_n$ hold true, then Conjecture C_n also holds true.*

Before we prove Theorem 6.2.5, we prepare two easy lemmas.

Lemma 6.2.6 *Let $f : X \to Y$ and $g : X \to Z$ be surjective morphisms between projective varieties. Let us consider*

$$U = \bigcap_{m \in \mathbb{Z}} U_m$$

such that U_m is a nonempty Zariski open set of Y for every $m \in \mathbb{Z}$. Then $f(g^{-1}(z)) \cap U \neq \emptyset$ for every sufficiently general point $z \in Z$.

Proof Since $f : X \to Y$ is surjective, $f^{-1}(U_m)$ is a nonempty Zariski open set of X. Since $g : X \to Z$ is surjective, $g(f^{-1}(U_m))$ is constructible by Chevalley's theorem and contains a nonempty Zariski open set V_m of Z (see, for example, [Har3, Chap. II, Exercise 3.19]). We put

$$V = \bigcap_{m \in \mathbb{Z}} V_m.$$

Then $f(g^{-1}(z)) \cap U \neq \emptyset$ for every $z \in V$ by construction. This is what we wanted. $\qquad \square$

Lemma 6.2.7 *Let $f : X \to Y$ and $g : X \to Z$ be surjective morphisms between normal projective varieties with connected fibers. Then there exist a general subvariety Γ of Z, a projective variety $\widehat{Y}$ and morphisms $\alpha : \widehat{Y} \to \Gamma$, $\beta : \widehat{Y} \to Y$ such that α is a surjective morphism with connected fibers, $\alpha^{-1}(z) \simeq f(g^{-1}(z))$ for $z \in \Gamma$, and $\beta : \widehat{Y} \to Y$ is dominant with $\dim \widehat{Y} = \dim Y$.*

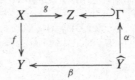

Proof of Lemma 6.2.7 We consider $\varphi = (f, g) : X \to Y \times Z$. Let W be the closure of $\operatorname{Im}\varphi$. By restricting the projection morphisms, we have $p : W \to Y$ and $q : W \to Z$.

$$
\begin{array}{ccc}
X & \xrightarrow{\ g\ } & Z \\
\downarrow{f} \ \ \searrow{\varphi} & & \uparrow{q} \\
Y & \xleftarrow[\ p\ ]{} & W
\end{array}
$$

For $z \in Z$, we have

$$q^{-1}(z) = (f(g^{-1}(z)), z) \simeq f(g^{-1}(z)),$$

and for $y \in Y$, we have

$$p^{-1}(y) = (y, g(f^{-1}(y))) \simeq g(f^{-1}(y)).$$

Therefore, p is surjective. We put $r = \dim W - \dim Y$. If $r = 0$, then we put $\Gamma = Z$ and $\widehat{Y} = W$. Assume that $r > 0$ holds. If $\dim Z \geq 2$, then we take a general hyperplane section Z_1 of Z. If Z is a curve, then we take a general point Z_1 of Z. In this case, $W_1 = q^{-1}(Z_1)$ is isomorphic to $W \times_Z Z_1$ and $W_1 = W \cap (Y \times Z_1)$. The map $p_1 = p|_{W_1} : W_1 \to Y$ satisfies that W_1 is a variety and

$$p_1^{-1}(y) = (y, g(f^{-1}(y)) \cap Z_1).$$

Since Z_1 is general, for a general point y of Y, it follows that

$$\dim(g(f^{-1}(y)) \cap Z_1) = r - 1.$$

By repeating this process, we have Z_r and W_r. Then $\Gamma = Z_r$ and $\widehat{Y} = W_r$ have the required property. □

Let us prove Theorem 6.2.5.

Proof of Theorem 6.2.5 We will use induction on n. We may assume that Conjecture C_k for $k \leq n - 1$, Conjectures $(\geq)_k$ and $(>)_k$ for $k \leq n$ hold true (see Remark 6.2.4). Let $f : X \to Y$ be a surjective morphism between smooth projective varieties with connected fibers such that $\dim X = n$. We have to prove the inequality

$$\kappa(X) \geq \kappa(X_y) + \kappa(Y), \tag{6.2.1}$$

where X_y is a sufficiently general fiber of $f : X \to Y$.

Step 1 By Conjecture $(\geq)_n$, we have $\kappa(X_y) = -\infty$ or $\kappa(Y) = -\infty$ if $\kappa(X) = -\infty$. Therefore, (6.2.1) clearly holds true when $\kappa(X) = -\infty$.

Step 2 We assume $\kappa(X) = 0$. If $\kappa(X_y) = -\infty$ or $\kappa(Y) = -\infty$, then (6.2.1) is obvious. Therefore, we further assume that $\kappa(X_y) \geq 0$ and $\kappa(Y) \geq 0$. By Conjecture $(>)_n$, $\kappa(X_y) = \kappa(Y) = 0$ holds when $\kappa(X) = 0$, $\kappa(X_y) \geq 0$, and $\kappa(Y) \geq 0$. In particular, (6.2.1) holds true in this case.

Thus, it is sufficient to prove the inequality under the assumption that $\kappa(X) > 0$, $\kappa(X_y) \geq 0$, and $\kappa(Y) \geq 0$.

Step 3 Let $g : X \to Z$ be the Iitaka fibration associated to X after modifying X birationally. For a sufficiently general point $z \in Z$, $X_z' = g^{-1}(z)$ has Kodaira dimension 0. Of course, $g : X \to Z$ is smooth over a neighborhood of $z \in Z$. Let us consider

$$f' = f|_{X_z} : X_z \to f(g^{-1}(z)).$$

By considering the Stein factorization

$$f' : X_z \longrightarrow V \xrightarrow{\pi} f(g^{-1}(z))$$

of this morphism, we get

$$0 = \kappa(X_z) \geq \kappa(X_{z,v}) + \kappa(V) \tag{6.2.2}$$

by induction on dimension, where $X_{z,v}$ is a sufficiently general fiber of $X_z \to V$. By Lemma 6.2.6, we may assume that $y = \pi(v)$ is a sufficiently general point of Y such that $f : X \to Y$ is smooth over a neighborhood of $y \in Y$. In particular, $X_y = g^{-1}(y)$ is smooth. By Lemma 6.2.7, we have Γ, $\widehat{Y}$, α, and β. We may assume that $z \in \Gamma$ by the construction of Γ in the proof of Lemma 6.2.7. By construction, we have $\alpha^{-1}(z) \simeq f(g^{-1}(z))$. Since $\dim Y = \dim \widehat{Y}$, $\kappa(\widehat{Y}) \geq \kappa(Y)$ follows. By Iitaka's easy addition formula (see Lemma 2.3.31), we get

$$\begin{aligned}
\kappa(\widehat{Y}) &\leq \kappa(\alpha^{-1}(z)) + \dim \Gamma \\
&\leq \kappa(V) + \dim Y - \dim V.
\end{aligned} \tag{6.2.3}$$

By hypothesis, $\kappa(Y) \geq 0$. Then

$$0 \leq \kappa(Y) \leq \kappa(\widehat{Y}) \leq \kappa(V) + \dim Y - \dim V.$$

This implies that $\kappa(V) \geq 0$. Since $y = \pi(v)$ is a sufficiently general point of $f(g^{-1}(z))$, $(f')^{-1}(y)$ is smooth, $g' = g|_{X_y} : X_y \to g(X_y)$ is flat over a neighborhood of $z \in g(X_y)$ (see [Mat, Theorem 22.5 and its corollary]), and $g(X_y)$ is smooth in a neighborhood of $z \in g(X_y)$ (see [Mat, Theorem 23.7]). By construction,

$$(f')^{-1}(y) = X_z \cap X_y = g^{-1}(z) \cap f^{-1}(y) = (g')^{-1}(z).$$

Thus we note that $X_{z,v}$ is a connected component of $(g')^{-1}(z)$. By taking the Stein factorization (see, for example, [FG, Lemma 2.4]) and using Iitaka's easy addition formula (see Lemma 2.3.31 and Remark 2.3.32), we have

$$\kappa(X_y) \leq \kappa(X_{z,v}) + \dim g(X_y). \tag{6.2.4}$$

By hypothesis, $\kappa(X_y) \geq 0$. Thus $\kappa(X_{z,v}) \geq 0$ follows. Therefore, (6.2.2) implies $\kappa(X_{z,v}) = \kappa(V) = 0$. Hence

$$\kappa(X_y) \leq \dim g(X_y)$$

by (6.2.4). Clearly, we have

$$\begin{aligned}
\kappa(X_y) \leq \dim g(X_y) &= \dim X_y - \dim X_{z,v} \\
&= \dim X - \dim Y - (\dim X_z - \dim V) \\
&= \dim Z + \dim V - \dim Y \\
&= \kappa(X) + \dim V - \dim Y.
\end{aligned}$$

Hence, $\kappa(X) \geq \kappa(X_y) + \dim Y - \dim V$. However, since $\kappa(V) = 0$, it follows that $\dim Y - \dim V \geq \kappa(\widehat{Y}) \geq \kappa(Y)$ by (6.2.3). Therefore, we get $\kappa(X) \geq \kappa(X_y) + \kappa(Y)$. This is what we wanted.

We finish the proof of Theorem 6.2.5. $\qquad\qquad\qquad\qquad\qquad\qquad\square$

References

[AbK]   D. Abramovich, K. Karu, Weak semistable reduction in characteristic 0. Invent. Math. **139**(2), 241–273 (2000)

[AbO]   D. Abramovich, F. Oort, Alterations and resolution of singularities, in *Resolution of Singularities (Obergurgl, 1997)*. Progress in Mathematics, vol. 181 (Birkhäuser, Basel, 2000), pp. 39–108

[AbTW]   D. Abramovich, M. Temkin, J. Włodarczyk, Relative desingularization and principalization of ideals, in preparation

[AdLT]   K. Adiprasito, G. Liu, M. Temkin, Semistable reduction in characteristic 0 (2018). arXiv:1810.03131 [math.AG]

[AlK]   A. Altman, S. Kleiman, *Introduction to Grothendieck Duality Theory*. Lecture Notes in Mathematics, vol. 146 (Springer, Berlin, 1970)

[BHPV]   W.P. Barth, K. Hulek, C.A.M. Peters, A. Van de Ven, *Compact Complex Surfaces*, 2nd edn. Ergebnisse der Mathematik und ihrer Grenzgebiete. 3. Folge. A Series of Modern Surveys in Mathematics, vol. 4 (Results in Mathematics and Related Areas. 3rd Series. A Series of Modern Surveys in Mathematics) (Springer, Berlin, 2004)

[BPV]   W.P. Barth, C. Peters, A. Van de Ven, *Compact Complex Surfaces*. Ergebnisse der Mathematik und ihrer Grenzgebiete (3), vol. 4 (Results in Mathematics and Related Areas (3)) (Springer, Berlin, 1984)

[Bi1]   C. Birkar, On existence of log minimal models II. J. Reine Angew. Math. **658**, 99–113 (2011)

[Bi2]   C. Birkar, Existence of flips and minimal models for 3-folds in char p. Ann. Sci. Éc. Norm. Supér. (4) **49**(1), 169–212 (2016)

[BCHM]   C. Birkar, P. Cascini, C.D. Hacon, J. McKernan, Existence of minimal models for varieties of log general type. J. Am. Math. Soc. **23**(2), 405–468 (2010)

[BCZ]   C. Birkar, Y. Chen, L. Zhang, Iitaka $C_{n,m}$ conjecture for 3-folds over finite fields. Nagoya Math. J. **229**, 21–51 (2018)

[BW]   C. Birkar, J. Waldron, Existence of Mori fibre spaces for 3-folds in char p. Adv. Math. **313**, 62–101 (2017)

[Cam]   F. Campana, Orbifolds, special varieties and classification theory. Ann. Inst. Fourier (Grenoble) **54**(3), 499–630 (2004)

[CaP]   J. Cao, M. Păun, Kodaira dimension of algebraic fiber spaces over abelian varieties. Invent. Math. **207**(1), 345–387 (2017)

© The Author(s), under exclusive license to Springer Nature Singapore Pte Ltd. 2020
O. Fujino, *Iitaka Conjecture*, SpringerBriefs in Mathematics,
https://doi.org/10.1007/978-981-15-3347-1

[ChZ]   Y. Chen, L. Zhang, The subadditivity of the Kodaira dimension for fibrations of relative dimension one in positive characteristics. Math. Res. Lett. **22**(3), 675–696 (2015)

[Co]   B. Conrad, *Grothendieck Duality and Base Change*. Lecture Notes in Mathematics, vol. 1750 (Springer, Berlin, 2000)

[dJ1]   A.J. de Jong, Smoothness, semi-stability and alterations. Inst. Hautes Études Sci. Publ. Math. **83**, 51–93 (1996)

[dJ2]   A.J. de Jong, Families of curves and alterations. Ann. Inst. Fourier (Grenoble) **47**(2), 599–621 (1997)

[DM]   Y. Dutta, T. Murayama, Effective generation and twisted weak positivity of direct images. Algebra Number Theory **13**(2), 425–454 (2019)

[Ec]   T. Eckl, Numerical analogues of the Kodaira dimension and the abundance conjecture. Manuscr. Math. **150**(3–4), 337–356 (2016)

[Ej]   S. Ejiri, Weak positivity theorem and Frobenius stable canonical rings of geometric generic fibers. J. Algebr. Geom. **26**(4), 691–734 (2017)

[EjZ]   S. Ejiri, L. Zhang, Iitaka's $C_{n,m}$ conjecture for 3-folds in positive characteristic. Math. Res. Lett. **25**(3), 783–802 (2018)

[Es]   H. Esnault, Classification des variétés de dimension 3 et plus (d'après T. Fujita, S. Iitaka, Y. Kawamata, K. Ueno, E. Viehweg), in *Bourbaki Seminar, Vol. 1980/81*. Lecture Notes in Mathematics, vol. 901 (Springer, Berlin, 1981), pp. 111–131

[EsV]   H. Esnault, E. Viehweg, *Lectures on Vanishing Theorems*. DMV Seminar, vol. 20 (Birkhäuser Verlag, Basel, 1992)

[F1]   O. Fujino, $\overline{C}_{n,n-1}$ revisited, preprint (2003)

[F2]   O. Fujino, Algebraic fiber spaces whose general fibers are of maximal Albanese dimension. Nagoya Math. J. **172**, 111–127 (2003)

[F3]   O. Fujino, A canonical bundle formula for certain algebraic fiber spaces and its applications. Nagoya Math. J. **172**, 129–171 (2003)

[F4]   O. Fujino, Higher direct images of log canonical divisors. J. Differ. Geom. **66**(3), 453–479 (2004)

[F5]   O. Fujino, Remarks on algebraic fiber spaces. J. Math. Kyoto Univ. **45**(4), 683–699 (2005)

[F6]   O. Fujino, What is log terminal? in *Flips for 3-Folds and 4-Folds*. Oxford Lecture Series in Mathematics and Its Applications, vol. 35 (Oxford University Press, Oxford, 2007), pp. 49–62

[F7]   O. Fujino, Fundamental theorems for the log minimal model program. Publ. Res. Inst. Math. Sci. **47**(3), 727–789 (2011)

[F8]   O. Fujino, Fundamental theorems for semi log canonical pairs. Algebr. Geom. **1**(2), 194–228 (2014)

[F9]   O. Fujino, On quasi-Albanese maps, preprint (2014)

[F10]   O. Fujino, Direct images of pluricanonical bundles. Algebr. Geom. **3**(1), 50–62 (2016)

[F11]   O. Fujino, Corrigendum: Direct images of relative pluricanonical bundles (Algebr. Geom. **3**(1), 50–62 (2016)), Algebr. Geom. **3**(2), 261–263 (2016)

[F12]   O. Fujino, *Foundations of the Minimal Model Program*. MSJ Memoirs, vol. 35 (Mathematical Society of Japan, Tokyo, 2017)

[F13]   O. Fujino, On subadditivity of the logarithmic Kodaira dimension. J. Math. Soc. Jpn. **69**(4), 1565–1581 (2017)

[F14]   O. Fujino, On semipositivity, injectivity and vanishing theorems, in *Hodge Theory and L^2-Analysis*. Advanced Lectures in Mathematics (ALM), vol. 39 (International Press, Somerville, 2017), pp. 245–282

[F15]   O. Fujino, Notes on the weak positivity theorems, in *Algebraic Varieties and Automorphism Groups*. Advanced Studies in Pure Mathematics, vol. 75 (Mathematical Society of Japan, Tokyo, 2017), pp. 73–118

[F16]   O. Fujino, Semipositivity theorems for moduli problems, Ann. Math. (2) **187**(3), 639–665 (2018)

[F17]   O. Fujino, Vanishing and semipositivity theorems for semi-log canonical pairs. Publ. Res. Inst. Math. Sci. **56**(1), 15–32 (2020)

[F18]   O. Fujino, Corrigendum to "On subadditivity of the logarithmic Kodaira dimension", to appear in J. Math. Soc. Jpn

[F19]   O. Fujino, On mixed-ω-sheaves (2019). arXiv:1908.00171 [math.AG]

[FF1]   O. Fujino, T. Fujisawa, Variations of mixed Hodge structure and semipositivity theorems. Publ. Res. Inst. Math. Sci. **50**(4), 589–661 (2014)

[FF2]   O. Fujino, T. Fujisawa, On semipositivity theorems. Math. Res. Lett. **26**(5), 1359–1382 (2019)

[FFS]   O. Fujino, T. Fujisawa, M. Saito, Some remarks on the semipositivity theorems. Publ. Res. Inst. Math. Sci. **50**(1), 85–112 (2014)

[FG]   O. Fujino, Y. Gongyo, On images of weak Fano manifolds. Math. Z. **270**(1–2), 531–544 (2012)

[FM]   O. Fujino, S. Mori, A canonical bundle formula. J. Differ. Geom. **56**(1), 167–188 (2000)

[Fs]   T. Fujisawa, A remark on semipositivity theorems (2017). arXiv:1710.01008 [math.AG]

[Ft1]   T. Fujita, Some remarks on Kodaira dimensions of fiber spaces. Proc. Jpn. Acad. Ser. A Math. Sci. **53**(1), 28–30 (1977)

[Ft2]   T. Fujita, Theory of the Kodaira dimension (its past, present, future). (Japanese) Sūgaku **30**(3), 243–254 (1978)

[G]   A. Grothendieck, Éléments de géométrie algébrique. IV. Étude locale des schémas et des morphismes de schémas. III, Inst. Hautes Études Sci. Publ. Math. (28), (1966)

[HPS]   C.D. Hacon, M. Popa, C. Schnell, Algebraic fiber spaces over abelian varieties: around a recent theorem by Cao and Păun, in *Local and Global Methods in Algebraic Geometry*. Contemporary Mathematics, vol. 712 (American Mathematical Society, Providence, 2018), pp. 143–195

[HX]   C.D. Hacon, C. Xu, On the three dimensional minimal model program in positive characteristic. J. Am. Math. Soc. **28**(3), 711–744 (2015)

[Har1] R. Hartshorne, Ample vector bundles. Inst. Hautes Études Sci. Publ. Math. **29**, 63–94 (1966)

[Har2] R. Hartshorne, *Residues and Duality*. Lecture Notes of a Seminar on the Work of A. Grothendieck, given at Harvard 1963/64. With an appendix by P. Deligne. Lecture Notes in Mathematics, vol. 20 (Springer, Berlin, 1966)

[Har3] R. Hartshorne, *Algebraic Geometry*. Graduate Texts in Mathematics, vol. 52 (Springer, New York, 1977)

[Har4] R. Hartshorne, Stable reflexive sheaves. Math. Ann. **254**(2), 121–176 (1980)

[Has1] K. Hashizume, On the non-vanishing conjecture and existence of log minimal models. Publ. Res. Inst. Math. Sci. **54**(1), 89–104 (2018)

[Has2] K. Hashizume, Log Iitaka conjecture for abundant log canonical fibrations (2019). arXiv:1902.10923 [math.AG]

[HH]   K. Hashizume, Z. Hu, On minimal model theory for log abundant lc pairs, to appear in J. Reine Angew. Math

[HNT]   K. Hashizume, Y. Nakamura, H. Tanaka, Minimal model program for log canonical threefolds in positive characteristic, to appear in Math. Res. Lett

[I1]   S. Iitaka, On D-dimensions of algebraic varieties. J. Math. Soc. Jpn. **23**, 356–373 (1971)

[I2]   S. Iitaka, Genus and classification of algebraic varieties. I. (Japanese) Sūgaku **24**(1), 14–27 (1972)

[I3]   S. Iitaka, On logarithmic Kodaira dimension of algebraic varieties, in *Complex Analysis and Algebraic Geometry* (Tokyo, Iwanami Shoten, 1977), pp. 175–189

[I4]   S. Iitaka, *Birational Geometry for Open Varieties*. Séminaire de Mathématiques Supérieures, vol. 76 (Presses de l'Université de Montréal, Montreal, 1981)

[I5]   S. Iitaka, *Algebraic Geometry. An Introduction to Birational Geometry of Algebraic Varieties*. Graduate Texts in Mathematics, vol. 76. North-Holland Mathematical Library, 24 (Springer, Berlin, 1982)

[I6]   S. Iitaka, Birational geometry and Kodaira dimension in various contexts. (Japanese)
            Sūgaku **34**(4), 289–300 (1982)

[Kar1]   K. Karu, Semistable reduction in characteristic zero. Boston University dissertation,
            1999

[Kar2]   K. Karu, Minimal models and boundedness of stable varieties. J. Algebr. Geom. **9**(1),
            93–109 (2000)

[KatM]   N. Katz, B. Mazur, *Arithmetic Moduli of Elliptic Curves*. Annals of Mathematics Stud-
            ies, vol. 108 (Princeton University Press, Princeton, 1985)

[Kaw1]   Y. Kawamata, On deformations of compactifiable complex manifolds. Math. Ann.
            235(3), 247–265 (1978)

[Kaw2]   Y. Kawamata, Addition formula of logarithmic Kodaira dimensions for morphisms of
            relative dimension one, in *Proceedings of the International Symposium on Algebraic
            Geometry (Kyoto University, Kyoto, 1977)* (Tokyo, Kinokuniya Book Store, 1978), pp.
            207–217

[Kaw3]   Y. Kawamata, Characterization of abelian varieties. Compos. Math. **43**(2), 253–276
            (1981)

[Kaw4]   Y. Kawamata, Kodaira dimension of algebraic fiber spaces over curves. Invent. Math.
            66(1), 57–71 (1982)

[Kaw5]   Y. Kawamata, Kodaira dimension of certain algebraic fiber spaces. J. Fac. Sci. Univ.
            Tokyo Sect. IA Math. **30**(1), 1–24 (1983)

[Kaw6]   Y. Kawamata, Minimal models and the Kodaira dimension of algebraic fiber spaces. J.
            Reine Angew. Math. **363**, 1–46 (1985)

[KKMS]   G. Kempf, F.F. Knudsen, D. Mumford, B. Saint-Donat, *Toroidal Embeddings. I*. Lecture
            Notes in Mathematics, vol. 339 (Springer, Berlin, 1973)

[Kol1]   J. Kollár, Higher direct images of dualizing sheaves. I. Ann. Math. (2) **123**(1), 11–42
            (1986)

[Kol2]   J. Kollár, Higher direct images of dualizing sheaves. II. Ann. Math. (2) **124**(1), 171–202
            (1986)

[Kol3]   J. Kollár, Subadditivity of the Kodaira dimension: fibers of general type, in *Alge-
            braic Geometry, Sendai, 1985*. Advanced Studies in Pure Mathematics, vol. 10 (North-
            Holland, Amsterdam, 1987), pp. 361–398

[Kol4]   J. Kollár, Projectivity of complete moduli. J. Differ. Geom. **32**(1), 235–268 (1990)

[Kol5]   J. Kollár, *Shafarevich Maps and Automorphic Forms* M. B. Porter Lectures (Princeton
            University Press, Princeton, 1995)

[KolM]   J. Kollár, S. Mori, *Birational Geometry of Algebraic Varieties*. With the collaboration
            of C.H. Clemens and A. Corti. Translated from the 1998 Japanese original. Cambridge
            Tracts in Mathematics, vol. 134 (Cambridge University Press, Cambridge, 1998)

[KovP]   S.J. Kovács, J. Patakfalvi, Projectivity of the moduli space of stable log-varieties and
            subadditivity of log-Kodaira dimension. J. Am. Math. Soc. **30**(4), 959–1021 (2017)

[Lai]   C.-J. Lai, Varieties fibered by good minimal models. Math. Ann. **350**(3), 533–547 (2011)

[Laz1]   R. Lazarsfeld, *Positivity in Algebraic Geometry. I. Classical Setting: Line Bundles and
            Linear Series*, Ergebnisse der Mathematik und ihrer Grenzgebiete. 3. Folge. A Series
            of Modern Surveys in Mathematics (Results in Mathematics and Related Areas. 3rd
            Series. A Series of Modern Surveys in Mathematics), vol. 48 (Springer, Berlin, 2004)

[Laz2]   R. Lazarsfeld, *Positivity in Algebraic Geometry. II. Positivity for Vector Bundles, and
            Multiplier Ideals*, Ergebnisse der Mathematik und ihrer Grenzgebiete. 3. Folge. A Series
            of Modern Surveys in Mathematics (Results in Mathematics and Related Areas. 3rd
            Series. A Series of Modern Surveys in Mathematics), vol. 49 (Springer, Berlin, 2004)

[Leh]   B. Lehmann, Comparing numerical dimensions. Algebra Number Theory **7**(5), 1065–
            1100 (2013)

[Les]   J. Lesieutre, Notions of numerical Iitaka dimension do not coincide (2019).
            arXiv:1904.10832 [math.AG]

[LD]   Z. Lu, M.R. Douglas, Gauss–Bonnet–Chern theorem on moduli space. Math. Ann.
            357(2), 469–511 (2013)

[Mae]   K. Maehara, The weak 1-positivity of direct image sheaves. J. Reine Angew. Math. **364**, 112–129 (1986)

[Mat]   H. Matsumura, *Commutative Ring Theory*. Translated from the Japanese by M. Reid. Cambridge Studies in Advanced Mathematics, vol. 8, 2nd edn. (Cambridge University Press, Cambridge, 1989)

[Mo1]   S. Mori, Threefolds whose canonical bundles are not numerically effective. Ann. Math. (2) **116**(1), 133–176 (1982)

[Mo2]   S. Mori, Classification of higher-dimensional varieties, in *Algebraic Geometry, Bowdoin, 1985* (Brunswick, Maine, 1985). Proceedings of Symposia in Pure Mathematics, vol. 46, Part 1. (American Mathematical Society, Providence, 1987)

[Mu]   D. Mumford, *Abelian Varieties*. With appendices by C.P. Ramanujam and Yuri Manin. Corrected reprint of the second (1974) edition. Tata Institute of Fundamental Research Studies in Mathematics, vol. 5. Published for the Tata Institute of Fundamental Research, Bombay (Hindustan Book Agency, New Delhi 2008)

[N1]   N. Nakayama, Invariance of the plurigenera of algebraic varieties under minimal model conjectures. Topology **25**(2), 237–251 (1986)

[N2]   N. Nakayama, Hodge filtrations and the higher direct images of canonical sheaves. Invent. Math. **85**(1), 217–221 (1986)

[N3]   N. Nakayama, Local structure of an elliptic fibration, in *Higher Dimensional Birational Geometry (Kyoto, 1997)*. Advanced Studies in Pure Mathematics, vol. 35 (Mathematical Society of Japan, Tokyo, 2002), pp. 185–295

[N4]   N. Nakayama, *Zariski-Decomposition and Abundance*. MSJ Memoirs, vol. 14 (Mathematical Society of Japan, Tokyo, 2004)

[Pat1]   Z. Patakfalvi, Semi-positivity in positive characteristics, Ann. Sci. Éc. Norm. Supér. (4) **47**(5), 991–1025 (2014)

[Pat2]   Z. Patakfalvi, On subadditivity of Kodaira dimension in positive characteristic over a general type base. J. Algebr. Geom. **27**(1), 21–53 (2018)

[Pău]   M. Păun, Siu's invariance of plurigenera: a one-tower proof. J. Differ. Geom. **76**(3), 485–493 (2007)

[PăT]   M. Păun, S. Takayama, Positivity of twisted relative pluricanonical bundles and their direct images. J. Algebr. Geom. **27**(2), 211–272 (2018)

[PopS]   M. Popa, C. Schnell, On direct images of pluricanonical bundles. Algebra Number Theory **8**(9), 2273–2295 (2014)

[Popp]   H. Popp, *Moduli Theory and Classification Theory of Algebraic Varieties*. Lecture Notes in Mathematics, vol. 620 (Springer, Berlin, 1977)

[Sh]   T. Shibata, On generic vanishing for pluricanonical bundles. Mich. Math. J. **65**(4), 873–888 (2016)

[Si]   Y.-T. Siu, Extension of twisted pluricanonical sections with plurisubharmonic weight and invariance of semipositively twisted plurigenera for manifolds not necessarily of general type, in *Complex Geometry (Göttingen, 2000)* (Springer, Berlin, 2002), pp. 223–277

[U1]   K. Ueno, *Classification Theory of Algebraic Varieties and Compact Complex Spaces*. Notes written in collaboration with P. Cherenack. Lecture Notes in Mathematics, vol. 439 (Springer, Berlin, 1975)

[U2]   K. Ueno, Bimeromorphic geometry of complex manifolds. (Japanese) Sūgaku **33**(3), 213–226 (1981)

[vaGO]   B. van Geemen, F. Oort, A compactification of a fine moduli space of curves, in *Resolution of Singularities (Obergurgl, 1997)*. Progress in Mathematics, vol. 181 (Birkhäuser, Basel, 2000), pp. 285–298

[Ve]   J.-L. Verdier, Base change for twisted inverse image of coherent sheaves, in *Algebraic Geometry (International Colloquium, Tata Institute of Fundamental Research, Bombay, 1968)* (Oxford University Press, London, 1969), pp. 393–408

[Vi1]   E. Viehweg, Canonical divisors and the additivity of the Kodaira dimension for morphisms of relative dimension one. Compos. Math. **35**(2), 197–223 (1977)

[Vi2]   E. Viehweg, Correction to: "Canonical divisors and the additivity of the Kodaira dimension for morphisms of relative dimension one" (Compos. Math. **35**(2), 197–223 (1977)). Compos. Math. **35**(3), 336 (1977)

[Vi3]   E. Viehweg, Weak positivity and the additivity of the Kodaira dimension for certain fibre spaces, in *Algebraic Varieties and Analytic Varieties (Tokyo, 1981)*. Advanced Studies in Pure Mathematics, vol. 1 (North-Holland, Amsterdam, 1983), pp. 329–353

[Vi4]   E. Viehweg, Weak positivity and the additivity of the Kodaira dimension. II. The local Torelli map, in *Classification of Algebraic and Analytic Manifolds (Katata, 1982)*. Progress in Mathematics, vol. 39 (Birkhäuser Boston, Boston, 1983), pp. 567–589

[Vi5]   E. Viehweg, Vanishing theorems and positivity in algebraic fibre spaces, in *Proceedings of the International Congress of Mathematicians*, Berkeley, California, vol. 1, 2 (American Mathematical Society, Providence, 1987), pp. 682–688

[Vi6]   E. Viehweg, Weak positivity and the stability of certain Hilbert points. Invent. Math. **96**(3), 639–667 (1989)

[Vi7]   E. Viehweg, *Quasi-Projective Moduli for Polarized Manifolds*. Ergebnisse der Mathematik und ihrer Grenzgebiete (3) (Results in Mathematics and Related Areas (3)), vol. 30 (Springer, Berlin, 1995)

[Wa1]   J. Waldron, Finite generation of the log canonical ring for 3-folds in char p. Math. Res. Lett. **24**(3), 933–946 (2017)

[Wa2]   J. Waldron, The LMMP for log canonical 3-folds in characteristic $p > 5$. Nagoya Math. J. **230**, 48–71 (2018)

[Z]   L. Zhang, Subadditivity of Kodaira dimensions for fibrations of three-folds in positive characteristics. Adv. Math. **354**, 106741 (2019), 29 pp

Index

© The Author(s), under exclusive license to Springer Nature Singapore Pte Ltd. 2020
O. Fujino, *Iitaka Conjecture*, SpringerBriefs in Mathematics,
https://doi.org/10.1007/978-981-15-3347-1

Printed in the United States
By Bookmasters